Lecture Notes in Mathematics

Edited by A. Dold, F. Takens and B. Teissier

Editorial Policy
for the publication of monographs

1. Lecture Notes aim to report new developments in ⸱⸱⸱⸱as of mathematics – quickly, informally and at a high level. Monograph manuscripts should be reasonably self-contained and rounded off. Thus they may, and often will, present not only results of the author but also related work by other people. They may be based on specialized lecture courses. Furthermore, the manuscripts should provide sufficient motivation, examples and applications. This clearly distinguishes Lecture Notes from journal articles or technical reports which normally are very concise. Articles intended for a journal but too long to be accepted by most journals, usually do not have this "lecture notes" character. For similar reasons it is unusual for doctoral theses to be accepted for the Lecture Notes series.

2. Manuscripts should be submitted (preferably in duplicate) either to one of the series editors or to Springer-Verlag, Heidelberg. In general, manuscripts will be sent out to 2 external referees for evaluation. If a decision cannot yet be reached on the basis of the first 2 reports, further referees may be contacted: the author will be informed of this. A final decision to publish can be made only on the basis of the complete manuscript, however a refereeing process leading to a preliminary decision can be based on a pre-final or incomplete manuscript. The strict minimum amount of material that will be considered should include a detailed outline describing the planned contents of each chapter, a bibliography and several sample chapters.
Authors should be aware that incomplete or insufficiently close to final manuscripts almost always result in longer refereeing times and nevertheless unclear referees' recommendations, making further refereeing of a final draft necessary.
Authors should also be aware that parallel submission of their manuscript to another publisher while under consideration for LNM will in general lead to immediate rejection.

3. Manuscripts should in general be submitted in English.
Final manuscripts should contain at least 100 pages of mathematical text and should include
- a table of contents;
- an informative introduction, with adequate motivation and perhaps some historical remarks: it should be accessible to a reader not intimately familiar with the topic treated;
- a subject index: as a rule this is genuinely helpful for the reader.

Lecture Notes in Mathematics

1710

Editors:
A. Dold, Heidelberg
F. Takens, Groningen
B. Teissier, Paris

Lecture Notes in Mathematics

Editors:
A. Dold
F. Takens
B. Teissier

1710

Springer
Berlin
Heidelberg
New York
Barcelona
Hong Kong
London
Milan
Paris
Singapore
Tokyo

Max Koecher

The Minnesota Notes
on Jordan Algebras
and Their Applications

Edited and annotated
by Aloys Krieg and Sebastian Walcher

Springer

Author

Max Koecher†

Editors

Aloys Krieg
Lehrstuhl A für Mathematik
RWTH Aachen
D-52056 Aachen, Germany
E-mail: krieg@mathA.rwth-aachen.de

Sebastian Walcher
Zentrum Mathematik
TU München
D-80290 München, Germany
E-mail: walcher@mathematik.tu-muenchen.de

Cataloging-in-Publication Data applied for

Die Deutsche Bibliothek – CIP-Einheitsaufnahme

Koecher, Max: The Minnesota notes on Jordan Algebras and their applications / Max Koecher.
Ed. and annot. by Aloys Krieg and Sebastian Walcher. – Berlin; Heidelberg; New York; Barcelona;
Hong Kong; London; Milan; Paris; Singapore; Tokyo: Springer, 1999
(Lecture notes in mathematics; 1710)
ISBN 3-540-66360-6

The material of this volume was originally published in 1962 under the title "Jordan
Algebras and Their Applications" by the School of Mathematics of the University of
Minnesota.

Mathematics Subject Classification (1991): 17C36, 17C55, 32M15, 32N10,
43A85, 17-02, 22-99

ISSN 0075-8434
ISBN 3-540-66360-0 Springer-Verlag Berlin Heidelberg New York

Typesetting: Camera-ready T$_E$X output by the editors

SPIN: 10650255   41/3143-543210 - Printed on acid-free paper

Preface

These notes contain the material of lectures on Jordan algebras given by me at the University of Minnesota during the academic year 1961-62. I tried to make use of the algebraic structure of a Jordan algebra over the field of real resp. complex numbers in certain applications to analysis. In the center of the investigatons stands the bijective correspondence between semisimple real Jordan algebras and the so-called ω-domains. This correspondence was known for domains of positivity. Most of the material is taken from an unpublished manuscript written jointly with E. Artin and H. Braun.

I wish to express my thanks to my friends and colleagues E. Calabi, H. Röhrl and L. Green for their continuous interest and their valuable discussions and suggestions. I am also grateful for H. Röhrl's help in preparing these notes.

Minneapolis, Minnesota, May 23, 1962   *Max Koecher*

Editors' Preface

For a number of years, Max Koecher's Minnesota lecture notes have been frequently quoted but have been, for all practical purposes, unavailable. Moreover, the few remaining mimeographed copies are almost illegible. Recently, interest in homogeneous symmetric cones has surged again, due to applications in fields as diverse as analysis and statistics. There exist new texts introducing the subject (notably the monograph [FKo] by Faraut and Korányi), but nevertheless there are, in the editors' opinion, several compelling reasons to re-issue Koecher's notes.

First, they comprise a classical reference that is still in demand, and should not be lost.

Second, the main objects of these notes are homogeneous, but not necessarily convex, cones, which have recently proven useful in applications.

Third, the notes contain an excellent introduction to the "Braun-Koecher view" of Jordan structures and their applications in analysis, with the additional advantage of being short and quickly accessible. The monograph [BK] by Braun and Koecher develops Jordan theory systematically and on a broad basis, but it is less suitable for obtaining a quick overview.

Naturally, these notes may be, in some respects, dated (like any text that approaches the age of forty), but they contain a clear and concise presentation of the material that does not, in our opinion, require any fundamental changes. There are minor modifications, most of them concerning typographical errors and language. At some points we felt that a change was necessary to clarify matters. The editors' notes at the end of each chapter contain comments, and an account of the relevant developments of the theory since these notes were first written. The author's references were updated whenever appropriate, and the editors included a number of new references that reflect the historical development since the original release of Koecher's notes. This list of additional references is not in any way complete, and several important publications are omitted. The editors did not intend to present a complete account of all the relevant books and articles that have been published since 1962. We hope, however, that his list will be a useful starting point for those who want to familiarize themselves with the theory and applications of Jordan structures. Readers interested in a general overview of recent developments may consult the conference proceedings [KMP] and [F-L].

Several colleagues have, over the past years, suggested a re-edition of these notes. The decisive initiative that convinced the editors to take up this work came from Jacques Faraut and Erhard Neher.

We would like to express our special thanks to Josef Dorfmeister, Simon Gindikin, Adam Korányi, Ottmar Loos, Holger Petersson and Dieter Pumplün, who sent us detailed and very helpful comments on preliminary versions. Moreover we are grateful to several colleagues, who have helped us in various ways, in particular Jacques Faraut, Antonio Fernandez-Lopez, Joachim Hilgert, Ulrich Hirzebruch, Wilhelm Kaup, Gerard Letac, Kurt Meyberg, Erhard Neher, Helmut Röhrl, Ichiro Satake, Audrey Terras, and Harald Upmeier. Last not least, we thank Birgit Morton for her careful preparation of the manuscript.

Aachen, June 1999   *Aloys Krieg, Sebastian Walcher*

Contents

Chapter I. Domains of Positivity

§1. Some notions and notations

Let K be a commutative field and X a vector space over K. We assume that the dimension of X over K is finite and denote it by n,

$$n := \dim_K X.$$

The elements of X will be denoted by $a, b, \ldots, x, y, z$ and the elements of the field K by $\alpha, \beta, \ldots, \xi, \omega$.

A map $l : X \to K$ is called a *linear form* on the space X if $l(\alpha x + \beta y) = \alpha l(x) + \beta l(y)$ holds for all $x, y \in X$ and $\alpha, \beta \in K$. The set X^* of all linear forms on X is a vector space over K in a natural way, via $(\alpha l_1 + \beta l_2)(x) := \alpha l_1(x) + \beta l_2(x)$. The dimension of X^* over K is equal to the dimension n of X over K. We call X^* the *dual space* of X.

It is well known that every $(n-1)$-dimensional subspace of X is given by an equation $l(x) = 0$ for a suitable, non-trivial linear form l. By a *hyperplane* of X we mean each subset of X given by an equation $l(x) = \alpha$ for a fixed non-trivial linear form $l \in X^*$ and $\alpha \in K$.

For each linear transformation A of X, i.e. for each endomorphism $A : X \to X$ of the vector space X, the *determinant* of A and the *trace* of A are well-defined. Indeed let $u_1, \ldots, u_n$ be a basis of X. Then the linear transformation A possesses a representation by a matrix $B = (\beta_{kl})$ given as follows

$$A\left(\sum_{k=1}^{n} \xi_k u_k\right) = \sum_{k=1}^{n}\left(\sum_{l=1}^{n} \beta_{kl}\xi_l\right) u_k.$$

Then we put

$$\det A := \det B := \det(\beta_{kl}) \quad \text{and} \quad \mathrm{tr}(A) := \mathrm{tr}(B) := \sum_{k=1}^{n} \beta_{kk}.$$

It is easily verified that the determinant $\det A$ and the trace $\mathrm{tr}(A)$ do not depend on the choice of the basis. We call the matrix B the *representation of*

A *with respect to the basis* $u_1, \ldots, u_n$. The linear transformation A has the kernel $\{0\}$ if and only if $\det A \neq 0$. In this case A is bijective and the inverse transformation will be denoted by A^{-1}.

A map $\sigma : X \times X \to K$ is a *symmetric bilinear form* if

(SBF.1) $\sigma(x, y) = \sigma(y, x)$, $x, y \in X$,

(SBF.2) $\sigma(x, \alpha y + \beta z) = \alpha \sigma(x, y) + \beta \sigma(x, z)$, $x, y, z \in X$, $\alpha, \beta \in K$.

σ is called *non-singular* if $\sigma(x, y) = 0$ for all $y \in X$ implies $x = 0$.

Let $\sigma(x, y)$ be a fixed non-singular symmetric bilinear form on X and $a \in X$. By the definition

$$a^* : X \to K, \ a^*(x) := \sigma(a, x),$$

we get a linear form a^* on X. Obviously the map

$$\varphi : X \to X^*, \ a \mapsto a^*,$$

is a homomorphism of vector spaces. The kernel of φ consists of all $a \in X$ such that $\sigma(a, x) = 0$ for all $x \in X$. Since σ is non-singular, it follows that $a = 0$. Hence the map φ is injective. Since X and X^* have the same finite dimension, it follows that φ is an isomorphism. Consequently, given a linear form $l \in X^*$ there exists a unique vector $l^* \in X$ such that

$$l(x) = \sigma(l^*, x), \quad x \in X.$$

It is obvious that $a^{**} = a$ and $l^{**} = l$ hold for all $a \in X$ and $l \in X^*$.

Let A be a linear transformation of X and $a \in X$. The map

$$X \to K, \ x \mapsto \sigma(a, Ax),$$

is a linear form. Consequently, there exists a unique vector $\tilde{a} \in X$ such that $\sigma(a, Ax) = \sigma(\tilde{a}, x)$, $x \in X$. Obviously the map

$$A^* : X \to X, \ a \mapsto \tilde{a},$$

is an endomorphism of X. We call A^* the *adjoint transformation* of A (with respect to σ), thus:

$$\sigma(y, Ax) = \sigma(A^*y, x), \quad x, y \in X.$$

Given linear transformations A, B of X, we have

$$(\alpha A + \beta B)^* = \alpha A^* + \beta B^*, \quad (AB)^* = B^* A^*, \quad (A^*)^* = A.$$

Let $\xi_1, \ldots, \xi_n$ resp. $\eta_1, \ldots, \eta_n$ be the coefficients of x resp. y with respect to a basis $u_1, \ldots, u_n$. Then the bilinear form σ is represented by a symmetric matrix $S = (\sigma_{kl}) = (\sigma(u_k, u_l))$ as follows:

$$\sigma(x,y) = \sum_{k,l=1}^{n} \xi_k \sigma_{kl} \eta_l.$$

Let A be a linear transformation of X with the matrix representation B with respect to $u_1, \ldots, u_n$. Then the matrix representation of A^* is given by

$$B^* = S^{-1} B^t S,$$

where B^t denotes the transpose of the matrix B. Consequently we have

$$\det A^* = \det A \quad \text{and} \quad \operatorname{tr}(A^*) = \operatorname{tr}(A).$$

A linear transformation A is called *self-adjoint* if $A^* = A$. In view of $\sigma(x, Ay) = \sigma(A^*x, y) = \sigma(y, A^*x)$ the bilinear form $\sigma_A(x,y) := \sigma(x, Ay)$ is symmetric if and only if $A^* = A$. On the other hand let $\tau(x,y)$ be an arbitrary symmetric bilinear form on X. Given $x \in X$ there exists a unique $\tilde{x} \in X$ satisfying $\tau(x,y) = \sigma(\tilde{x},y)$, $y \in X$, and $A : X \to X$, $x \mapsto \tilde{x}$, is a linear transformation. It follows that

$$\tau(x,y) = \sigma(Ax, y), \quad x,y \in X.$$

Since τ is symmetric, A is self-adjoint. Summarizing there exists a bijective correspondence between the self-adjoint linear transformations of X and the symmetric bilinear forms on X.

Let $K = \mathbf{R}$ be the field of real numbers. Using an isomorphism between X and $\mathbf{R}^n$, the vector space X possesses a natural topology. With respect to this topology the maps

$$X \to X, \qquad x \mapsto \alpha x, \qquad \alpha \in \mathbf{R},$$
$$X \to \mathbf{R}, \qquad x \mapsto l(x), \qquad l \in X^*,$$
$$X \times X \to \mathbf{R}, (x,y) \mapsto \sigma(x,y),$$

are continuous. This natural topology on X can be given by an arbitrary *norm* on X, i.e. a map $X \to \mathbf{R}$, $x \mapsto |x|$, which has the following properties

$$|x| > 0 \quad \text{for} \quad x \in X, \ x \neq 0, \ |0| = 0,$$
$$|\lambda x| = |\lambda| \cdot |x|, \ \lambda \in \mathbf{R}, \quad x \in X,$$
$$|x + y| \leq |x| + |y|, \quad x,y \in X.$$

Let $Hom(X,X)$ denote the set of all linear transformations of X. Then $Hom(X,X)$ turns out to be a vector space over $\mathbf{R}$ of dimension n^2 and therefore one has a natural topology on $Hom(X,X)$. On the other hand we can define a norm on $Hom(X,X)$ as follows: Putting

$$|W| := \sup\left\{ \frac{|Wx|}{|x|}; \ 0 \neq x \in X \right\} = \sup\{|Wx|; \ x \in X, \ |x| = 1\}$$

for $X \neq \{0\}$, then $|\cdot|$ becomes a norm on $Hom(X, X)$, satisfying

$$|W_1 W_2| \leq |W_1| \cdot |W_2|, \quad W_1, W_2 \in Hom(X, X).$$

This norm also defines the natural topology on $Hom(X, X)$. Especially, the set $\{W \in Hom(X, X); |W| \leq \gamma\}$ is compact in $Hom(X, X)$.

Denote by $GL(X)$ the general linear group of X, i.e. the group of all $W \in Hom(X, X)$ such that $\det W \neq 0$. Thus $GL(X)$ is an open subset of $Hom(X, X)$ and is a locally compact topological group with respect to the induced topology.

We have seen that the set $\{W \in Hom(X, X); |W| \leq \gamma\}$ is compact in $Hom(X, X)$. The set $\{W \in GL(X); |W| \leq \gamma\}, \gamma > 0$, is obviously *not* compact in $GL(X)$. Now we consider the set

$$K := \{W \in GL(X); |W| \leq \gamma_1, |W^{-1}| \leq \gamma_2\}.$$

Since the map $W \mapsto |\det W|$ is continuous and $|W^{-1}| \leq \gamma_2$, the absolute value of the determinant of W^{-1} is bounded. In view of

$$1 = |\det W| \cdot |\det W^{-1}| \leq \gamma_3 |\det W|,$$

K is closed in $GL(X)$. On the other hand, K is contained in the intersection of $GL(X)$ and the compact set $\{W \in Hom(X, X); |W| \leq \gamma_1\}$. Therefore K is compact in $GL(X)$.

A self-adjoint linear transformation A of X is called *positive definite* if

$$\sigma(Ax, x) > 0 \quad \text{for all} \quad x \in X, \ x \neq 0.$$

A is called *positive semidefinite* if

$$\sigma(Ax, x) \geq 0 \quad \text{for all} \quad x \in X.$$

The fixed bilinear form σ is said to be *positive definite* if the identity transformation is positive definite, that means if

$$\sigma(x, x) > 0 \quad \text{for all} \quad x \in X, \ x \neq 0.$$

A positive definite bilinear form σ defines a *volume element* on the vector space X. Indeed, let $\tau_1, \ldots, \tau_n$ be the components of the point $t \in X$ with respect to some basis $u_1, \ldots, u_n$ and let $S = (\sigma_{kl})$ the matrix of the bilinear form σ. Then we have

$$dt = |\det S|^{-1/2} d\tau_1 \ldots d\tau_n.$$

It is easy to see that the volume element dt is independent of the choice of the basis.

If G is an open and relative compact subset of X, and $f : \overline{G} \to \mathbf{C}$ is continuous, one has the (Riemann) *integral*

$$\int_G f(t)dt$$

As is well-known, this notion can be extended to include functions satisfying certain growth conditions on unbounded open domains.

The notions *differentiable, real-analytic, rational, polynomial*, etc., may be defined in coordinates as soon as a basis is given, and are independent of the choice of the basis.

§2. The notion of a domain of positivity

For the rest of this chapter let X be a vector space over $\mathbf{R}$ of dimension n. We assume that $|x|$ is a norm on X, which therefore defines the natural topology on X. Moreover let $\sigma(x,y)$ be a non-singular symmetric bilinear form on X.

Definition. A subset Y of X is called a *domain of positivity* (with respect to σ) if the following three axioms are fulfilled:

(P.0) Y is open and non-empty.

(P.1) Given $a, b \in Y$ one has $\sigma(a, b) > 0$.

(P.2) If $x \in X$ then $\sigma(a, x) > 0$ for all $a \in \overline{Y}$, $a \neq 0$, implies $x \in Y$.

$\overline{Y}$ stands for the closure and ∂Y for the boundary of Y in the natural topology on X. It is easy to see that these three axioms are independent.

Since the bilinear form $\sigma(x,y)$ is continuous we conclude that

$$\sigma(x,y) \geq 0 \quad \text{for} \quad x,y \in \overline{Y}.$$

More precisely we have

Theorem 1. *Given a domain of positivity Y and $x \in X$ the following hold:*

(a) $x \in Y$ is equivalent to $\sigma(a, x) > 0$ for all $0 \neq a \in \overline{Y}$.

(b) $x \in \overline{Y}$ is equivalent to $\sigma(a, x) \geq 0$ for all $a \in Y$.

Proof. (a) "$\Rightarrow$": Given $x \in Y$ we have

$$\sigma(a, x) \geq 0 \quad \text{for} \quad a \in \overline{Y}.$$

For fixed $a \neq 0$ the map $x \mapsto \sigma(a, x)$ is a non-trivial linear form on X. Hence the map is open. Consequently, the image of the open set Y is open in $\mathbf{R}$, thus $\sigma(a, x) > 0$ for $x \in Y$.

"$\Leftarrow$": That is (P.2)

(b) "$\Rightarrow$": This follows from (P.1) by virtue of continuity.

"$\Leftarrow$": Let be $\sigma(a, x) \geq 0$ for all $a \in Y$. Then $\sigma(a, x) \geq 0$ holds for all $a \in \overline{Y}$. For $a \neq 0$ and fixed $b \in Y$ we have $\sigma(a, b) > 0$ by (a). Consequently, for $\mu > 0$ we get $\sigma(a, x + \mu b) = \sigma(a, x) + \mu\sigma(a, b) > 0$. This is true for each $a \in \overline{Y}$, $a \neq 0$. Thus (P.2) implies $x + \mu b \in Y$. By $\mu \to 0$ we obtain $x \in \overline{Y}$. $\square$

It follows directly from this theorem that $\overset{\circ}{\overline{Y}} = Y$; i.e. the open kernel of $\overline{Y}$ is equal to Y.

Theorem 2. *Let Y be a domain of positivity. Then one has*

(a) If $x \in \overline{Y}$ and $y \in Y$ then $x + y \in Y$.

(b) If $x \in Y$ and $\lambda > 0$ then $\lambda x \in Y$.

(c) Given $0 \neq a \in \partial Y$ there exists $0 \neq b \in \partial Y$ satisfying $\sigma(a, b) = 0$.

(d) If $a \in \overline{Y}$ and $-a \in \overline{Y}$ then $a = 0$.

(e) H is a support plane of Y if and only if $H = \{x \in X;\ \sigma(a, x) = 0\}$ for some $0 \neq a \in \overline{Y}$.

A hyperplane H in X is called a *support plane* of a set $A \subset X$ if H can be described by an equation $\sigma(a, x) - \lambda = 0$ for some $a \in X$, $a \neq 0$, such that $\sigma(a, x) - \lambda \geq 0$ holds for all $x \in A$ and $\sigma(a, x) - \lambda = 0$ for at least one $x \in \overline{A}$.

The statements (a) and (b) imply that Y is a *convex cone*, i.e.

$$\alpha x + (1 - \alpha)y \in Y, \quad \beta x \in Y \quad \text{for all} \quad x, y \in Y,\ 0 \leq \alpha \leq 1,\ \beta > 0.$$

Moreover, it follows from (d) that Y is a *proper* convex cone, i.e. $\overline{Y}$ does not contain a one-dimensional subspace of X.

Proof. (a) Given $0 \neq a \in \overline{Y}$ we have $\sigma(a, x) \geq 0$ and $\sigma(a, y) > 0$ by Theorem 1. Consequently $\sigma(a, x + y) > 0$ and therefore $x + y \in Y$.

(b) Use Theorem 1(a).

(c) We have $\sigma(a, b) \geq 0$ for all $b \in \overline{Y}$, $b \neq 0$. If $\sigma(a, b) > 0$ for all $b \in \overline{Y}$, $b \neq 0$, then $a \in Y$ by (P.2). Consequently, there exists $b \in \overline{Y}$, $b \neq 0$, satisfying $\sigma(a, b) = 0$. Theorem 1(a) shows $b \notin Y$, hence $b \in \partial Y$.

(d) By assumption we have $\sigma(a, x) \geq 0$ and $-\sigma(a, x) \geq 0$ and thus $\sigma(a, x) = 0$ for all $x \in Y$. As Y is open, there exist n linearly independent vectors in Y, and therefore we have $\sigma(a, x) = 0$ for all $x \in X$. Since σ is non-singular, $a = 0$ follows.

(e) "$\Rightarrow$": Let $a \in X$, $a \neq 0$, $\lambda \in \mathbf{R}$ such that

$$H = \{x \in X; \ \sigma(a, x) - \lambda = 0\}$$

is a support plane of Y. We conclude that all $x \in \overline{Y}$ satisfy $\sigma(a, x) - \lambda \geq 0$. For $x = 0$ we obtain $\lambda \leq 0$. For $x = \mu y$, $\mu > 0$, $y \in \overline{Y}$, we have $\sigma(a, y) - \frac{1}{\mu}\lambda \geq 0$. Now $\mu \to \infty$ shows that $\sigma(a, y) \geq 0$ for all $y \in \overline{Y}$. Since H is a support plane, it follows that $\lambda = 0$. Now we apply Theorem 1(b) and obtain $a \in \overline{Y}$.

"$\Leftarrow$": Given $0 \neq a \in \overline{Y}$ then $\sigma(a, x) > 0$ holds for $x \in Y$ due to Theorem 1(a). In view of $0 \in \overline{Y}$ and $\sigma(a, 0) = 0$ the hyperplane $H := \{x \in X; \ \sigma(a, x) = 0\}$ is a support plane of Y. $\qquad \square$

Domains of positivity are *maximal* in the following sense: Let $Y_1 \subset Y_2$ be two domains of positivity with respect to the same bilinear form, then $Y_1 = Y_2$. Indeed, let $0 \neq y_1 \in \overline{Y_1}$ and $y_2 \in Y_2$. In view of $Y_1 \subset Y_2$ we have $0 \neq y_1 \in \overline{Y_2}$ and consequently $\sigma(y_1, y_2) > 0$ due to Theorem 1(a) for Y_2. The axiom (P.2) for Y_1 gives $y_2 \in Y_1$ and hence $Y_2 \subset Y_1$.

The following lemma is easy to prove but is useful:

Lemma 1. *Given any compact subset K in Y, there exists a positive number $\varrho = \varrho(K)$ such that*

$$\sigma(a, y) \geq \varrho(K) \cdot |y| \quad \text{for all} \quad a \in K \quad \text{and} \quad y \in \overline{Y}.$$

Proof. The set

$$K' := \{(a, y); \ a \in K, \ y \in \overline{Y}, \ |y| = 1\}$$

is compact in $\overline{Y} \times \overline{Y}$. Since $\sigma(a, y)$ is continuous, there exists $(a_0, y_0) \in K'$ such that $\sigma(a, y) \geq \sigma(a_0, y_0) > 0$ for $(a, y) \in K'$. Set $\varrho(K) := \sigma(a_0, y_0)$, then we obtain $\sigma(a, y) \geq \varrho(K)$ for $a \in K$, $y \in \overline{Y}$, $|y| = 1$. Homogeneity leads to

$$\sigma(a, y) \geq \varrho(K) \cdot |y| \quad \text{for} \quad a \in K, \ 0 \neq y \in \overline{Y}.$$

For $y = 0$ the assertion is obvious. $\qquad \square$

If K contains only a single point $a \in Y$ then we set $\varrho(K) =: \varrho(a)$ and have

$$\sigma(a,y) \geq \varrho(a) \cdot |y| \quad \text{for} \quad y \in \overline{Y}.$$

Recall that an *ordering* "$\geq$" on the set X satisfies

(0) $$x \geq x \quad \text{for all} \quad x \in X,$$

(1) $$x \geq y, \, y \geq z \quad \Rightarrow \quad x \geq z,$$

(2) $$x \geq y, \, y \geq x \quad \Rightarrow \quad x = y.$$

Due to Theorem 2, the definition

$$x \geq y \quad \Leftrightarrow \quad x - y \in \overline{Y}$$

gives an ordering. Moreover, this ordering is compatible with the vector space structure, i.e. we have

(3) $$x \geq y, \, \lambda > 0 \quad \Rightarrow \quad \lambda x \geq \lambda y,$$

(4) $$x \geq y, \, a \in X \quad \Rightarrow \quad a + x \geq a + y.$$

The bilinear form $\sigma(a, x)$ is monotone:

(5) $$a \geq 0, \, x \geq y \quad \Rightarrow \quad \sigma(a,x) \geq \sigma(a,y)$$

and "$\geq$" is an Archimedian ordering:

(6)   Given $a \in Y$, $x \in X$ there exists $\lambda > 0$ such that $\lambda a \geq x$.

Since the set $\{x \in X; \, |x| = 1\}$ is compact, we have $|\sigma(a,x)| \leq \kappa \cdot |a| \cdot |x|$ for all $a \in Y$, $x \in X$ (Schwarz inequality) with a certain positive κ. On the other hand, $\sigma(a,b) \geq \varrho(a) \cdot |b|$ holds for $b \in \overline{Y}$ by Lemma 1. Let $b \in Y$ and $\lambda > 0$. Then $\sigma(b, \lambda a - x) = \lambda \sigma(a,b) - \sigma(b,x) \geq \lambda \varrho(a) \cdot |b| - \kappa \cdot |b| \cdot |x| \geq |b| \cdot (\lambda \varrho(a) - \kappa |x|)$. For all λ, $\lambda > \kappa \cdot |x| / \varrho(a)$ we have $\sigma(b, \lambda a - x) \geq 0$ and Theorem 1(b) yields $\lambda a - x \geq 0$.

(7)   Given $a_1, \ldots, a_m \in Y$ there exists $d \in Y$ such that $a_\mu \geq d$, $1 \leq \mu \leq m$.

Indeed, let c be a fixed element in Y, and $\lambda_\mu > 0$ such that $\lambda_\mu a_\mu \geq c$ for $1 \leq \mu \leq m$. Hence, the claim (7) is true for $d := \min \left\{ \frac{1}{\lambda_1}, \ldots, \frac{1}{\lambda_m} \right\} \cdot c$.

(8)   There exists a positive γ_1 such that $x \geq y \geq 0$ implies $|x| \geq \gamma_1 \cdot |y|$.

In order to see this, let $d \in Y$. Then we have by (5), the Schwarz inequality, and Lemma 1 that $\kappa \cdot |x| \cdot |d| \geq \sigma(d,x) \geq \sigma(d,y) \geq \varrho(d) \cdot |y|$. The claim has been proved with $\gamma_1 := \varrho(d)/(\kappa \cdot |d|)$.

(9)   Given $a, b \in Y$, $a \le b$, the set $\{x \in X; a \le x \le b\}$ is a compact subset of Y.

Indeed, this set is closed in X and it follows from (8) that it is bounded. Therefore it is compact in X. In view of $y \ge a > 0$ the set is contained in Y.

(10)   Given $x, y \in \overline{Y}$, $y \ne 0$ there exists $\alpha \ge 0$ such that $x - \alpha y \in \partial Y$.

If $x \in \partial Y$ let $\alpha = 0$. Let $x \in Y$ and $x \ge x - \beta y \ge 0, \beta \ge 0$. Then (8) yields $|x| \ge \gamma_1 |x - \beta y| \ge \gamma \cdot (\beta |y| - |x|)$. Thus the set of possible β is bounded. Since $\overline{Y}$ is closed, there exists $\alpha \ge 0$ such that $x - \alpha y \in \overline{Y}$ and $x - \beta y \notin \overline{Y}$ for $\beta > \alpha$. We conclude $x - \alpha y \in \partial Y$, because Y is open.

(11)   If $n > 1$ one has $\{y_1 + y_2; y_1, y_2 \in \partial Y\} = \overline{Y}$.

The inclusion "⊂" follows from $y_1 + y_2 \ge y_1 + 0 \ge 0$ due to (4). As $n > 1$ there exist two linearly independent vectors in Y. Thus $\partial Y \ne \{0\}$ is a consequence of (10). Given $x \in \overline{Y}$ choose $0 \ne y \in \partial Y$ and $\alpha \ge 0$ with $x - \alpha y \in \partial Y$. The claim follows from $x = (x - \alpha y) + \alpha y$ and $\alpha y \in \partial Y$.

Besides the ordering "$\ge$" we introduce the relation "$>$" by

$$x > y \quad \Leftrightarrow \quad x - y \in Y.$$

It is easy to see that (1) and (3) to (8) remain valid if we replace "$\ge$" by "$>$". One should keep in mind that $x > y$ is not equivalent with $x \ge y$ and $x \ne y$.

§3. The automorphisms of a domain of positivity

A linear transformation $W : X \rightarrow X$ is called a (linear) *automorphism* of the domain of positivity Y if

$$W\overline{Y} = \overline{Y}.$$

Since Y is open and non-empty, it contains a basis of X. Hence the linear transformation W is non-singular, i.e. $\det W \ne 0$. The elements of $GL(X)$ are homeomorphisms of X. In view of $\overset{\circ}{\overline{Y}} = Y$ a linear transformation $W : X \rightarrow X$ therefore is an automorphism of Y if and only if

$$WY = Y.$$

In view of $\det W \ne 0$ the set of all automorphisms of Y is a (multiplicative) group $\Sigma(Y)$, a subgroup of $GL(X)$, the *automorphism group* of Y. Since Y is a cone, the map $x \mapsto \lambda x$, $\lambda > 0$, belongs to $\Sigma(Y)$.

Considering the bilinear form $\sigma(x, y)$, we have the adjoint transformation W^* of W.

Lemma 2. *If $W \in \Sigma(Y)$, then $W^* \in \Sigma(Y)$.*

Proof. Let $a \in \overline{Y}$ and $y \in Y$. Then one has $Wy \in Y$ and therefore $\sigma(W^*a, y) = \sigma(a, Wy) \geq 0$. Theorem 1(b) leads to $W^*a \in \overline{Y}$ or $W^*\overline{Y} \subset \overline{Y}$. By virtue of $W^{-1} \in \Sigma(Y)$ and $(W^*)^{-1} = (W^{-1})^*$ we also have $W^{*-1}\overline{Y} \subset \overline{Y}$. Consequently $W^*\overline{Y} = \overline{Y}$ follows. □

In view of $(W_1 W_2)^* = W_2^* W_1^*$ the map

$$\Sigma(Y) \to \Sigma(Y), \ W \mapsto W^*,$$

is an antiautomorphism. $\Sigma(Y)$ is a subgroup of $GL(X)$, and therefore a topological group. Obviously, the map $W \mapsto W^*$ is a homeomorphism. $\Sigma(Y)$ is a closed set in the induced topology on $GL(X)$. Indeed, let $A \in GL(X)$, $A = \lim_{\mu \to \infty} W_\mu$, $W_\mu \in \Sigma(Y)$. In view of $\det A \neq 0$, it follows that also $\lim_{\mu \to \infty} W_\mu^{-1} = A^{-1}$ holds. We have $W_\mu \overline{Y} = \overline{Y}$ and $W_\mu^{-1}\overline{Y} = \overline{Y}$, and consequently $A\overline{Y} \subset \overline{Y}, A^{-1}\overline{Y} \subset \overline{Y}$, or $A\overline{Y} = \overline{Y}$. This yields $A \in \Sigma(Y)$.

Now we recall the concept of a transformation group. Let Y be a topological space and Σ a topological group. Σ is said to be a *transformation group on Y* if

(T.1) Given $W \in \Sigma$ and $x \in Y$ then $Wx \in Y$.

(T.2) The map $\Sigma \times Y \to Y, (W, x) \mapsto Wx$, is continuous.

Σ is called a *proper transformation group* if moreover

(T.3) The map $\Phi : \Sigma \times Y \to Y \times Y, (W, x) \mapsto (x, Wx)$, is *proper*, i.e. the inverse image of each compact subset is compact.

Since Φ can be written as the composition $(W, x) \mapsto (x, W, x) \mapsto (x, Wx)$, it is clearly continuous.

Theorem 3. *Given a domain of positivity Y, the automorphism group $\Sigma(Y)$ is a proper transformation group on Y.*

Proof. We only need to prove that for every compact subset K of $Y \times Y$, the inverse image $\Phi^{-1}(K)$ is compact in $\Sigma \times Y$. We may assume that K has the form $K = K_1 \times K_2$, where K_1, K_2 are compact subsets of Y, because every compact subset of $Y \times Y$ is contained in a compact set of this type. Since Φ is continuous, the inverse image $\Phi^{-1}(K_1 \times K_2)$ is closed in $\Sigma \times Y$. Hence, the theorem is proved if we can find a compact subset M of $\Sigma(Y)$ such that

(1) $$\Phi^{-1}(K_1 \times K_2) \subset M \times K_1.$$

Let

$$M := M(K_1, K_2) := \{W \in \Sigma(Y);\ WK_1 \cap K_2 \neq \emptyset\}.$$

Then (1) is fulfilled. Indeed, given $(W, x) \in \Phi^{-1}(K_1 \times K_2)$, then $\Phi(W, x) = (x, Wx) \in K_1 \times K_2$, hence $x \in K_1$ and $Wx \in K_2$. Therefore we have $Wx \in WK_1 \cap K_2$ and $(W, x) \in M \times K_1$.

Now we show that M is compact. Given $W \in M$ there exists $x_W \in K_1$ such that $Wx_W \in K_2$. For $y \in Y$ we apply Lemma 1 and obtain

$$\sigma(Wx_W, y) = \sigma(x_W, W^*y) \geq \varrho(K_1) \cdot |W^*y|.$$

The Schwarz inequality $\sigma(a, b) \leq \kappa \cdot |a| \cdot |b|$ leads to

$$\sigma(Wx_W, y) \leq \kappa \cdot |Wx_W| \cdot |y| \leq \gamma_1 \cdot |y|,$$

because $Wx_W \in K_2$ and $|x|$ is bounded for $x \in K_2$. Summarizing we have

$$|W^*y| \leq \gamma_2 \cdot |y| \quad \text{for} \quad y \in Y$$

Since Y is open and therefore contains a basis of X, we conclude

$$|W^*x| \leq \gamma_3 \cdot |x| \quad \text{for} \quad x \in X.$$

By the definition of the norm $|W|$ we get

$$|W^*| \leq \gamma_3.$$

Since $W \mapsto W^*$ is a homeomorphism, we also have

(2) $$W \in M(K_1, K_2) \quad \Rightarrow \quad |W| \leq \gamma_4.$$

The definition of $M(K_1, K_2)$ leads to

$$W \in M(K_1, K_2) \quad \Rightarrow \quad W^{-1} \in M(K_2, K_1).$$

It follows from (2) that

$$W \in M(K_1, K_2) \quad \Rightarrow \quad |W| \leq \gamma_4, |W^{-1}| \leq \gamma_5.$$

Note that the γ_i depend only on K_1 and K_2.

From §1 we know that $M(K_1, K_2)$ is contained in a compact subset of $GL(X)$, and obviously, $M(K_1, K_2)$ is closed in $GL(X)$. Since $M(K_1, K_2) \subset \Sigma(Y)$ and since $\Sigma(Y)$ is closed in $GL(X)$, the set $M(K_1, K_2)$ is closed in $\Sigma(Y)$ and therefore compact.   □

Corollary 1. *Given $a \in Y$ then $\{W \in \Sigma(Y);\ Wa = a\}$ is a compact subgroup of $\Sigma(Y)$.*

Proof. Put $K_1 = K_2 = \{a\}$, then $\Phi^{-1}(K_1 \times K_2) = \{(W, a); \ Wa = a\}$. $\square$

Lemma 3. *Given a domain of positivity Y (with respect to σ), the following statements are equivalent:*

(i) Y is a domain of positivity with respect to τ.

(ii) There exists a self-adjoint $W \in \Sigma(Y)$ such that $\tau(x, y) = \sigma(Wx, y)$.

Proof. "(i) $\Rightarrow$ (ii)": Since τ is symmetric, there exists a self-adjoint linear transformation W such that $\tau(x, y) = \sigma(Wx, y)$. Given $y \in \overline{Y}$, $y \neq 0$ and $x \in Y$ therefore $\sigma(Wx, y) > 0$ holds. Theorem 1 leads to $Wx \in Y$ and $WY \subset Y$. Interchanging τ and σ we get $W^{-1}Y \subset Y$ and $W \in \Sigma(Y)$.

"(ii) $\Rightarrow$ (i)": Apply the definition as well as $WY = Y$ and $W\overline{Y} = \overline{Y}$. $\square$

§4. Norms of a domain of positivity

Let Y be a fixed domain of positivity with respect to $\sigma(x, y)$. A function $\varphi : \overline{Y} \to \mathbf{R}$ is said to be a *norm on Y* if

(N.1) φ is real-analytic and positive on Y, continuous on $\overline{Y}$.

(N.2) $\varphi(y) = 0$ for $y \in \partial Y$.

(N.3) $\varphi(Wy) = |\det W| \cdot \varphi(y)$ for $W \in \Sigma(Y)$ and $y \in Y$.

Since the maps $x \mapsto \lambda x, \lambda > 0$, belong to $\Sigma(Y)$, we have $\varphi(\lambda y) = \lambda^n \cdot \varphi(y)$. We will prove that there exists a norm on Y.

Let us consider the n–dimensional integral

$$\tilde{\omega}(y) := \int_Y e^{-\sigma(y, t)} dt, \quad y \in Y.$$

Lemma 1 shows that for a compact subset $K \subset Y$ we have

$$\int_Y e^{-\sigma(y, t)} dt \leq \int_Y e^{-\varrho(K) \cdot |t|} dt, \quad y \in K.$$

In view of

$$\int_Y e^{-\varrho(K) \cdot |t|} dt \leq \int_X e^{-\varrho(K) \cdot |t|} dt < \infty,$$

the integral for $\tilde{\omega}(y)$ is uniformly and absolutely convergent for $y \in K$. Consequently, $\tilde{\omega}(y)$ is real-analytic and positive on Y.

In the integral

$$\tilde{\omega}(Wy) = \int_Y e^{-\sigma(y, W^* t)} dt, \quad W \in \Sigma(Y),$$

we substitute $t = W^{*-1}x$. Thus $dt = |\det W^{*-1}| dx = |\det W|^{-1} dx$ leads to

$$(1) \qquad \tilde{\omega}(Wy) = |\det W|^{-1} \cdot \tilde{\omega}(y).$$

Theorem 4. *Given a positive constant ω_0 the function*

$$\omega : \overline{Y} \to \mathbf{R}, \; \omega(y) := \begin{cases} \omega_0/\tilde{\omega}(y) & \text{for} \;\; y \in Y, \\ 0 & \text{for} \;\; y \in \partial Y, \end{cases}$$

is a norm on Y satisfying

$$\omega(a) > \omega(b) \quad \text{for} \quad a > b > 0.$$

Proof. Continuity of $\omega(y)$ only remains to be proved on ∂Y. Therefore we need to show that $\tilde{\omega}(y)$ tends to infinity if $y \in Y$ tends to a point $a \in \partial Y$.

$n = 1$: Let $X = \mathbf{R}$, $\sigma(x, y) = \sigma \cdot x \cdot y$, $\sigma > 0$. Then we have $Y =]0; \infty[$, $\partial Y = \{0\}$ and $\tilde{\omega}(y) = \int_0^\infty e^{-\sigma \cdot y \cdot t} dt = \frac{1}{y} \int_0^\infty e^{-\sigma \cdot t} dt$. Thus the claim is obvious.

$n > 1$: There exists $0 \neq b \in \partial Y$ such that $\sigma(a, b) = 0$. Indeed, for $a = 0$ we may take any $b \in \partial Y \backslash \{0\}$, and for $a \neq 0$ we apply Theorem 2(c). Now we choose a compact set K_0 with $\overset{\circ}{K_0} \neq \emptyset$ (contained in some neighborhood of b) such that

(i) $K_0 \subset Y$,

(ii) $0 < \sigma(a, t) < \frac{1}{2}$ for $t \in K_0$.

Let

$$K_\lambda := K_0 + \lambda b = \{x \in X; \; x - \lambda b \in K_0\}.$$

Then we have

(i) $K_\lambda \subset Y$ for $\lambda > 0$,

(ii) $0 < \sigma(a, t) < \frac{1}{2}$ for $t \in K_\lambda$,

(iii) vol $K_\lambda =$ vol $K_0 > 0$.

Given $\gamma > 0$ there exists a disjoint union K of a finitely many sets K_λ, $\lambda > 0$, such that

$$0 < \sigma(a, t) < \frac{1}{2} \quad \text{for} \quad t \in K, \quad \text{vol } K > \gamma.$$

K is compact. Thus there exists an open neighborhood U of a satisfying

$$\sigma(y,t) < 1 \quad \text{for} \quad y \in U \quad \text{and} \quad t \in K.$$

Consequently we have for $y \in U \cap Y$

$$\tilde{\omega}(y) = \int_Y e^{-\sigma(y,t)} dt \geq \frac{1}{e} \int_K dt > \frac{\gamma}{e}.$$

Hence, $\tilde{\omega}(y) \to \infty \quad$ for $\quad y \to a, \ y \in Y$.

Finally by §2, (5), we have

$$\tilde{\omega}(a) < \tilde{\omega}(b) \quad \text{for} \quad a > b > 0.$$

Hence the theorem has been proved completely. $\square$

§5. Examples

(A) *Direct products.*

Let $X_\alpha, 1 \leq \alpha \leq k$, be vector spaces over $\mathbf{R}$, let $\sigma_\alpha(x_\alpha, y_\alpha)$ be symmetric non-singular bilinear forms on X_α and let Y_α be domains of positivity in X_α (with respect to σ_α). We put

$$
\begin{aligned}
X : &= X_1 \oplus X_2 \oplus \ldots \oplus X_k \\
&= \{x = x_1 + x_2 + \ldots + x_k; \ x_\alpha \in X_\alpha, \alpha = 1, \ldots, k\}, \\
\sigma(x,y) : &= \sigma_1(x_1, y_1) + \sigma_2(x_2, y_2) + \ldots + \sigma_k(x_k, y_k) \quad \text{for} \quad x, y \in X, \\
Y : &= \{y = y_1 + \ldots + y_k; \ y_\alpha \in Y_\alpha, \ \alpha = 1, \ldots, k\}.
\end{aligned}
$$

Then it is easy to see that Y is a domain of positivity (with respect to σ) in X. Considering the norm $\omega(y)$ we get

$$\omega(y) = \omega_0 \cdot \left(\int_Y e^{-\sigma(y,t)} dt \right)^{-1} = \omega_0 \cdot \prod_{\alpha=1}^k \left(\int_{Y_\alpha} e^{-\sigma_\alpha(y_\alpha, t_\alpha)} dt_\alpha \right)^{-1}$$

$$= \gamma \cdot \prod_{\alpha=1}^k \omega_\alpha(y_\alpha)$$

with some constant $\gamma > 0$.

In particular, if we start with the domains

$$X_\alpha = \mathbf{R}, \quad Y_\alpha = \{y \in \mathbf{R}; \ y > 0\}, \quad \sigma_\alpha(x,y) = x \cdot y,$$

their product in $\mathbf{R}^k$ turns out to be the domain of positivity

$$Y = \{y \in \mathbf{R}^k; \ y_\alpha > 0, \ \alpha = 1, \ldots, k\}, \ \sigma(x,y) = \sum_{\alpha=1}^k x_\alpha y_\alpha.$$

(B) *Positive definite quadratic forms.*

Let X be the set of all real symmetric $m \times m$ matrices $x = (\xi_{\nu\mu}) = (\xi_{\mu\nu})$, $n = m(m+1)/2$, and

$$\sigma(x,y) = \text{tr}(xy) = \sum_{\nu,\mu=1}^{m} \xi_{\nu\mu}\eta_{\mu\nu} \quad , \quad |x| = \sqrt{\sigma(x,x)}.$$

An $m \times m$ matrix $r = (\varrho_{\nu\mu})$ defines a linear transformation $W(r)$ on X by

$$x \mapsto W(r)x := rxr^t,$$

where r^t stands for the transpose of r. One easily verifies

(1) $$W(r_1 r_2) = W(r_1)W(r_2),$$

(2) $$W^*(r) = W(r^t).$$

It follows from (1) that $\psi(r) := |\det W(r)|$ satisfies

(3) $$\psi(r_1 r_2) = \psi(r_1)\psi(r_2).$$

A well-known theorem states that each continuous function $\psi(r)$ satisfying (3) and $\psi(r) > 0$ for $\det r \neq 0$ is a power of $|\det r|$. We will give a simple proof of this theorem in §6 under the hypothesis that $\psi(r)$ is differentiable. Hence we have $\psi(r) = |\det r|^\varrho$. Since the elements of $W(r)$ are quadratic polynomials in the components of r, we conclude

$$\psi(\lambda r) = |\det W(\lambda r)| = |\det \lambda^2 W(r)| = \lambda^{2n}|\det W(r)| = \lambda^{2n}\psi(r),$$

and consequently $\varrho = \frac{2n}{m} = m + 1$, i.e.

(4) $$|\det W(r)| = |\det r|^{m+1}.$$

A matrix $y = (\eta_{\nu\mu}) \in X$ is called *positive definite* if the quadratic form

$$\sum_{\nu,\mu=1}^{m} \eta_{\nu\mu}\xi_{\nu}\xi_{\mu}$$

is positive definite, which is equivalent to the property that all the eigenvalues of y are positive. Now we put

$$Y := \{y \in X; \ y \text{ positive definite }\}.$$

Obviously, we have

$$r \quad \text{real} \quad m \times m \quad \text{matrix}, \quad \det r \neq 0 \quad \Rightarrow \quad W(r) \in \Sigma(Y).$$

We prove that Y is a domain of positivity with respect to σ. Indeed, Y is open in X. In order to demonstrate (P.1) and (P.2) we apply the following well-known

Lemma 4. *Given $x \in X$ there exists a real matrix r with $\det r \neq 0$ such that $W(r)x$ is a diagonal matrix.*

Let $x, y \in X$ and

$$x = W(r)d, \quad d = (\delta_{\nu\mu}d_\nu) \quad \text{a diagonal matrix}, \quad y = W^{*-1}(r)a.$$

We get

$$\sigma(x, y) = \sigma(d, a) = \sum_{\nu=1}^{m} d_\nu a_{\nu\nu}.$$

For $x, y \in Y$, we have $a, d \in Y$ and therefore $\sigma(x, y) > 0$, because the diagonal entries of a positive definite matrix are positive. Let $x \in X$ and $\sigma(x, y) > 0$ for all $0 \neq y \in \overline{Y}$. We obtain

$$\sum_{\nu=1}^{m} d_\nu a_{\nu\nu} > 0 \quad \text{for} \quad 0 \neq a \in \overline{Y}$$

and therefore $d_\nu > 0$ and $d \in Y$, hence $x \in Y$.

Now, we consider a norm on Y: By definition we have

$$\varphi(Wy) = |\det W| \cdot \varphi(y).$$

For $W = W(r)$ we get

$$\varphi(ryr^t) = |\det r|^{m+1} \cdot \varphi(y).$$

Choosing r such that ryr^t is the unit matrix we have $\det y \cdot (\det r)^2 = 1$. Hence we obtain

$$\varphi(y) = \varphi_0 \cdot |\det r|^{-(m+1)} = \varphi_0 \cdot (\det y)^{(m+1)/2}.$$

Each norm on Y is therefore given by

$$\varphi_0 \cdot (\det y)^{(m+1)/2}.$$

(C) *The circular cone*

Let X a vector space over $\mathbf{R}$ and $\sigma(x, y)$ a positive definite symmetric bilinear form on X. We fix a point $c \in X$ such that $\sigma(c, c) = 1$. Putting

$$\mu(x, y) := 2\sigma(c, x) \cdot \sigma(c, y) - \sigma(x, y)$$

we obtain a new bilinear form μ on X. The bilinear form μ is also non-singular, since

$$0 = \mu(a, x) = 2\sigma(c, a) \cdot \sigma(c, x) - \sigma(a, x)$$

for all $x \in X$ implies $2\sigma(c, a)c = a$, hence $2\sigma(c, a) = \sigma(c, a)$ and $\sigma(c, a) = 0$. Therefore we have $0 = \mu(a, a) = -\sigma(a, a)$ and $a = 0$.

$$Y = \{y \in X;\ \mu(c, y) > 0,\ \mu(y, y) > 0\}$$

is open and non-empty in view of $c \in Y$. Fixing an orthogonal basis with respect to σ, which contains c, it is easy to prove that Y is a domain of positivity with respect to σ. Using this basis we can easily find a self-adjoint and orthogonal (with respect to σ) linear transformation $T \in Hom(X, X)$ such that

$$\mu(x, y) = \sigma(Tx, y) = \sigma(x, Ty), \quad \text{and} \quad Tc = c.$$

The definition of Y yields $TY = Y$, hence $T \in \Sigma(Y)$. Thus the set Y is also a domain of positivity with respect to the bilinear form $\mu(x, y)$ by virtue of Lemma 3. The special norm ω on Y (cf. Theorem 4) can be computed as

$$\omega(y) = \gamma_0 \cdot [\mu(y, y)]^{n/2}.$$

§6. Differential operators

Let X be a vector space over $\mathbf{R}$ of dimension n. Recall that, given an open subset D of X and a map

$f : D \to X$ (vector-valued function) resp. $f : D \to \mathbf{R}$ (scalar-valued function)

we have the notions of "rational", "real-analytic", "differentiable" maps, etc. for f. For a real interval I and a function $F : I \to X$ let

$$\dot{F}(\tau) = \frac{d}{d\tau} F(\tau) = \lim_{\varepsilon \to 0} \frac{F(\tau + \varepsilon) - F(\tau)}{\varepsilon} \quad (\in X),$$

provided that this limit exists.

For continuously differentiable

$$f : D \to X \quad \text{resp.} \quad f : D \to \mathbf{R}$$

and $u \in X$ we define the operator Δ_x^u by

$$\Delta_x^u f(x) := \frac{d}{d\tau} f(x + \tau u)\Big|_{\tau=0}.$$

Fixing a basis of X we obtain

$$\Delta_x^u f(x) = \sum_{\nu=1}^n \frac{\partial f(x)}{\partial \xi_\nu} \omega_\nu,$$

if $\xi_1, \ldots, \xi_n$ resp. $\omega_1, \ldots, \omega_n$ are the coordinates of x resp. u with respect to this basis. Consequently, given x, then $\Delta_x^u f(x)$ is linear in u. The Taylor formula says that

$$f(x + \tau u) = f(x) + \tau \cdot a + O(\tau^2)$$

for twice continuously differentiable f. (Here and in the following O denotes the Landau symbol.) Applying the operator Δ_x^u, we get $\Delta_x^u f(x) = a$, or

$$f(x + \tau u) = f(x) + \tau \cdot \Delta_x^u f(x) + O(\tau^2).$$

Now let f be a vector-valued and g a scalar- or vector-valued function. Then we have

$$g(f(x + \tau u)) = g\left(f(x) + \tau \Delta_x^u f(x) + O(\tau^2)\right)$$
$$= g(f(x)) + \tau \cdot \Delta_{f(x)}^a g(f(x)) + O(\tau^2), \quad a = \Delta_x^u f(x),$$

and consequently

(1) $$\Delta_x^u g(f(x)) = \Delta_{f(x)}^a g(f(x)), \quad a = \Delta_x^u f(x).$$

Since $\Delta_x^u f(x)$ is linear in u, the map $u \mapsto \Delta_x^u f(x)$ (for fixed x) is a linear transformation on X. Denoting this transformation by $\frac{\partial f(x)}{\partial x}$, we obtain

$$\Delta_x^u f(x) = \frac{\partial f(x)}{\partial x} u.$$

For vector-valued functions f and g it follows from (1) that

(2) $$\frac{\partial g(f(x))}{\partial x} = \frac{\partial g(f(x))}{\partial f(x)} \cdot \frac{\partial f(x)}{\partial x}.$$

For scalar-valued functions f and g, we have the product rule

$$\Delta_x^u f(x) g(x) = f(x) \Delta_x^u g(x) + g(x) \Delta_x^u f(x).$$

Now we assume that a symmetric non-singular bilinear form $\sigma(x, y)$ on X is given. We have seen that, for a given scalar-valued function g, the map $X \to \mathbf{R}, u \mapsto \Delta_x^u g(x)$, is linear. Consequently, this is a linear form on X, which is denoted by $\frac{\partial g(x)}{\partial x}$. By definition we have

$$\Delta_x^u g(x) = \frac{\partial g(x)}{\partial x} u.$$

For this linear form there exists a unique vector

$$\left(\frac{\partial g(x)}{\partial x}\right)^* \in X$$

such that

$$\Delta_x^u g(x) = \frac{\partial g(x)}{\partial x} u = \sigma\left(\left(\frac{\partial g(x)}{\partial x}\right)^*, u\right).$$

We call $\left(\frac{\partial g}{\partial x}\right)^*$ the *gradient* of g with respect to σ. If g is a scalar-valued and f a vector-valued function, then (1) leads to

$$\sigma\left(\left(\frac{\partial g(f(x))}{\partial x}\right)^*, u\right) = \Delta_x^u g(f(x)) = \Delta_{f(x)}^a g(f(x)) = \sigma\left(\left(\frac{\partial g(f(x))}{\partial f(x)}\right)^*, a\right)$$

$$= \sigma\left(\left(\frac{\partial g(f(x))}{\partial f(x)}\right)^*, \frac{\partial f(x)}{\partial x} u\right)$$

and consequently

$$\left(\frac{\partial g(f(x))}{\partial x}\right)^* = \left(\frac{\partial f(x)}{\partial x}\right)^* \left(\frac{\partial g(f(x))}{\partial f(x)}\right)^*.$$

Let f be a twice continuously differentiable scalar-valued function (defined on a domain). Then

$$\tau(u, v) = \Delta_x^u \Delta_x^v f(x)$$

is a symmetric bilinear form on X, which depends on $x \in X$ and f. Consequently, there exists a self-adjoint linear transformation $H(x)$ such that

$$\Delta_x^u \Delta_x^v f(x) = \sigma(H(x)u, v).$$

We find

$$\Delta_x^v f(x) = \sigma\left(\left(\frac{\partial f}{\partial x}\right)^*, v\right),$$

$$\Delta_x^u \Delta_x^v f(x) = \sigma\left(\Delta_x^u \left(\frac{\partial f}{\partial x}\right)^*, v\right) = \sigma\left(\frac{\partial}{\partial x}\left(\frac{\partial f}{\partial x}\right)^* u, v\right)$$

and therefore

$$H(x) = \frac{\partial}{\partial x}\left(\frac{\partial f}{\partial x}\right)^*.$$

Using the Taylor formula for $\varphi(\tau) = f(x + \tau u)$, we obtain

$$f(x + \tau u) = f(x) + \tau \frac{d}{d\lambda} \varphi(\lambda)\Big|_{\lambda=0} + \frac{1}{2}\tau^2 \frac{d^2}{d\lambda^2} \varphi(\lambda)\Big|_{\lambda=\Theta}, \quad 0 \leq \Theta \leq \tau,$$

provided that the whole line segment from x to u belongs to the domain. Therefore one has

$$f(x + \tau u) = f(x) + \tau \Delta_x^u f(x) + \frac{1}{2}\tau^2 \sigma(H(x + \Theta u)u, u), \quad 0 \leq \Theta \leq \tau.$$

We will later use the Euler differential equation for a homogeneous function. Let $f(x)$ be a scalar- or vector-valued differentiable function and suppose that for all $\tau > 0$, $f(\tau x) = \tau^\kappa f(x)$, with some $\kappa \in \mathbf{R}$. We get $\Delta_x^x f(x) = \frac{d}{d\tau} f(\tau x)\Big|_{\tau=1} = \kappa f(x)$ and therefore we have

$$\frac{\partial f(x)}{\partial x} x = \kappa \cdot f(x).$$

Let us consider an application: Let X be the set of all real $m \times m$ matrices. X is a vector space over $\mathbf{R}$ of dimension $n = m^2$. Putting

$$\sigma(x, y) = \operatorname{tr}(xy),$$

we obtain a non-singular symmetric bilinear form on X. With respect to this bilinear form we get the identities

$$\sigma(xy, z) = \sigma(x, yz) = \sigma(zx, y) \quad \text{for} \quad x, y, z \in X.$$

Now we consider a function $\psi : X \to \mathbf{R}$ satisfying:

(a) $\psi(x) > 0$ for $x \in X$ with $\det x \neq 0$.

(b) $\psi(x)$ is differentiable on X.

(c) $\psi(xy) = \psi(x) \cdot \psi(y)$ for $x, y \in X$.

Given $u \in X$ then $\Delta_x^u \log \psi(x)$ is a linear form in u and thus there exists a matrix $x^\sharp \in X$ such that

$$\sigma(u, x^\sharp) = \Delta_x^u \log \psi(x).$$

Applying Δ_x^u to (c) and using formula (2) we obtain

$$\sigma(u, x^\sharp) = \Delta_x^u \log \psi(xy) = \Delta_{xy}^a \log \psi(xy); \quad a = \Delta_x^u xy = uy;$$
$$= \Delta_{xy}^{uy} \log \psi(xy) = \sigma(uy, (xy)^\sharp) = \sigma(u, y(xy)^\sharp).$$

Therefore we get

(3) $$x^\sharp = y(xy)^\sharp.$$

Applying Δ_y^u to (c) the same arguments yield

(4)
$$y^\sharp = (xy)^\sharp x.$$

In (3) we put $y = x^{-1}$ and in (4) we substitute x by x^{-1} and put $y = x$. If e denotes the unit matrix, and $c = e^\sharp$, then

$$x^\sharp = x^{-1} \cdot c, \quad x^\sharp = c \cdot x^{-1}.$$

Therefore we have $x^{-1}c = cx^{-1}$, and hence $yc = cy$ for all $y \in X$. It follows that $c = \lambda \cdot e$ and

$$x^\sharp = \lambda x^{-1}.$$

Now we consider $\Delta_x^u \log \det x$ for $\det x > 0$ and obtain

$$\det(x + \tau u) = \det[x(e + \tau x^{-1}u)] = \det x \cdot \det(e + \tau x^{-1}u)$$
$$= (\det x) \cdot (1 + \tau \operatorname{tr}(x^{-1}u) + O(\tau^2)).$$

Hence, we have

$$\Delta_x^u \log \det x = \operatorname{tr}(x^{-1}u) = \sigma(u, x^{-1})$$

and

$$\Delta_x^u \log \frac{\psi(x)}{(\det x)^\lambda} = 0 \quad \text{for all} \quad u, x.$$

Therefore we get $\psi(x) = \kappa(\det x)^\lambda$ for some constant κ . Thus (c) implies

$$\psi(x) = |\det x|^\lambda.$$

The case $\det x < 0$ is treated similarly. Every solution of $(a) - (c)$ is therefore a power of the absolute value of the determinant of x.

§7. An invariant line element

Let Y be a domain of positivity with respect to $\sigma(x, y)$ and let $\omega(y)$ be a special norm on Y (see Theorem 4):

$$\omega(y) = \frac{\omega_0}{\tilde\omega(y)}, \quad \tilde\omega(y) = \int_Y e^{-\sigma(y,t)} dt.$$

Given $y \in Y, u, v \in X$, we put

$$\tau_y(u, v) := \frac{1}{2\tilde\omega^2(y)} \int_Y \int_Y e^{-\sigma(y, t_1 + t_2)} \sigma(u, t_1 - t_2) \sigma(v, t_1 - t_2) dt_2 dt_1.$$

Using Lemma 1 it is easy to see that the double integral is absolutely convergent. $\tau_y(u,v)$ is a symmetric bilinear form on X and $\tau_y(u,u)$ is positive for $u \neq 0$. Given $W \in \Sigma(Y)$ we have

$$\tau_{Wy}(Wu, Wv) = \frac{|\det W|^2}{2\tilde{\omega}^2(y)} \int_Y \int_Y e^{-\sigma(y, W^*(t_1+t_2))}$$
$$\cdot \sigma(u, W^*(t_1 - t_2))\, \sigma(v, W^*(t_1 - t_2))\, dt_2 dt_1.$$

The substitutions $t_1 = W^{*-1} x_1$, $t_2 = W^{*-1} x_2$, $dt_1 dt_2 = |\det W|^{-2} dx_1 dx_2$ therefore lead to

$$(1) \qquad\qquad \tau_{Wy}(Wu, Wv) = \tau_y(u, v).$$

There is another way to obtain $\tau_y(u,v)$. We have

$$\Delta_y^v \tilde{\omega}(y) = -\int_Y e^{-\sigma(y,t)} \sigma(v,t) dt,$$

$$\Delta_y^u \Delta_y^v \tilde{\omega}(y) = \int_Y e^{-\sigma(y,t)} \sigma(u,t) \sigma(v,t) dt,$$

$$\Delta_y^u \Delta_y^v \log \tilde{\omega}(y) = \Delta_y^u \frac{1}{\tilde{\omega}(y)} \Delta_y^v \tilde{\omega}(y) = \frac{1}{\tilde{\omega}^2} \left(\tilde{\omega} \Delta_y^u \Delta_y^v \tilde{\omega} - \Delta_y^u \tilde{\omega} \cdot \Delta_y^v \tilde{\omega} \right),$$

and therefore

$$\Delta_y^u \Delta_y^v \log \tilde{\omega}(y) = \frac{1}{\tilde{\omega}^2} \Big(\int_Y \int_Y e^{-\sigma(y,t_1)-\sigma(y,t_2)} \sigma(u,t_2)\sigma(v,t_2) dt_2 dt_1$$
$$- \int_Y \int_Y e^{-\sigma(y,t_1)-\sigma(y,t_2)} \sigma(u,t_1)\sigma(v,t_2) dt_2 dt_1 \Big).$$

By symmetrization we get

$$2\tilde{\omega}^2 \Delta_y^u \Delta_y^v \log \tilde{\omega}(y) = \int_Y \int_Y e^{-\sigma(y,t_1+t_2)}[\sigma(u,t_1)\sigma(v,t_1) + \sigma(u,t_2)\sigma(v,t_2)$$
$$-\sigma(u,t_1)\sigma(v,t_2) - \sigma(v,t_1)\sigma(u,t_2)] dt_2 dt_1$$
$$= \int_Y \int_Y e^{-\sigma(y,t_1+t_2)} \sigma(u,t_1 - t_2)\sigma(v,t_1 - t_2) dt_2 dt_1$$

and consequently

$$(2) \qquad\qquad \tau_y(u,v) = \Delta_y^u \Delta_y^v \log \tilde{\omega}(y) = -\Delta_y^u \Delta_y^v \log \omega(y).$$

Let

$$H(y) := -\frac{\partial}{\partial y}\left(\frac{\partial \log \omega(y)}{\partial y} \right)^*$$

for $y \in Y$. Then we know from §6 that

(3) $$T_y(u, v) = \sigma(H(y)u, v).$$

Since $T_y(u, v)$ is positive definite,

(4) $$H(y) \quad \text{is positive definite.}$$

Equation (1) implies

(5) $$W^* H(Wy)W = H(y), \quad W \in \Sigma(Y).$$

Let $y = y(\tau)$ be a curve in Y and set

$$ds = \sqrt{\sigma(H(y)\dot{y}, \dot{y})}d\tau.$$

Then we obtain a positive definite line element ds, which is invariant with respect to the mappings

$$Y \to Y, \quad y \mapsto Wy, \quad W \in \Sigma(Y).$$

§8. The map $y \mapsto y^{\sharp}$

Starting with a domain of positivity Y and a special norm $\omega(y)$ (cf. Theorem 4), we know that $\Delta_y^u \log \omega(y)$, $y \in Y$, is a linear form in u, hence of the form $\sigma(y^{\sharp}, u)$, $y^{\sharp} \in X$, i.e.

$$y^{\sharp} := \left(\frac{\partial}{\partial y} \log \omega(y)\right)^*, \quad y \in Y.$$

Since $\omega(y)$ is real-analytic, the mapping $Y \to X$, $y \mapsto y^{\sharp}$, is real-analytic.

Lemma 5. *(a) If $W \in \Sigma(Y)$ then*

$$(Wy)^{\sharp} = W^{*-1}y^{\sharp} \quad \text{for every} \quad y \in Y.$$

(b) The map $Y \to X$, $y \mapsto y^{\sharp}$, is continuous and open.

Proof. (a) We have

$$y^{\sharp} = \left(\frac{\partial}{\partial y} \log \omega(Wy)\right)^* = \left(\frac{\partial Wy}{\partial y}\right)^* \left(\frac{\partial \log \omega(Wy)}{\partial Wy}\right)^* = W^*(Wy)^{\sharp}.$$

(b) Equation (4) in §7 says that the linear transformation

$$H(y) = -\frac{\partial}{\partial y}\left(\frac{\partial \log \omega(y)}{\partial y}\right)^*$$

is positive definite. Hence the determinant $\det H(y)$ is different from 0. In view of $-H(y) = \frac{\partial y^{\sharp}}{\partial y}$, therefore the map $y \mapsto y^{\sharp}$ is open. □

Since the map $y \mapsto \lambda y, \lambda > 0$, belongs to $\Sigma(Y)$, it follows that

(1)
$$(\lambda y)^{\sharp} = \frac{1}{\lambda} y^{\sharp}.$$

Now we apply the Taylor formula to $\log \omega(y)$. We have for $x, x + u \in Y$

$$\log \omega(x + u) = \log \omega(x) + \Delta_x^u \log \omega(x) + \frac{1}{2} \Delta_z^u \Delta_z^u \log \omega(z)\big|_{z=x+\Theta u},$$

or

(2) $\log \omega(x + u) = \log \omega(x) + \sigma(x^{\sharp}, u) - \frac{1}{2}\sigma(H(x + \Theta u)u, u), \quad 0 \leq \Theta \leq 1.$

Since $H(x)$ is positive definite, it follows that

$$\log \omega(b) < \log \omega(a) + \sigma(a^{\sharp}, b - a), \quad a, b \in Y, \ a \neq b,$$

or

(3)
$$\frac{\omega(b)}{\omega(a)} < e^{\sigma(a^{\sharp}, b-a)}, \quad a, b \in Y, \quad a \neq b.$$

Given $a \in Y$

$$N_a := \{y \in Y; \ \omega(y) = \omega(a)\}$$

is called the *norm surface* through a. In view of $\omega(\lambda y) = \lambda^n \omega(y)$ we see that

(NS.1) For $y \in Y$, there exists one and only one $\lambda > 0$ such that $\lambda y \in N_a$.

(NS.2) N_a is connected.

Two arbitrary points $y, z \in N_a$ may be connected by the curve

$$\tau \mapsto \sqrt[n]{\frac{\omega(y)}{\omega(y + \tau(z - y))}} \cdot (y + \tau(z - y)), \quad 0 \leq \tau \leq 1.$$

Now we demonstrate

Theorem 5. *The map $Y \to Y, y \mapsto y^{\sharp}$, is a homeomorphism.*

Theorem 6. *(a) Each norm surface N_a is convex with respect to the origin.*
(b) Each $y \in N_a$, $y \neq a$, satisfies

$$\sigma(a^{\sharp}, y - a) > 0$$

and $\sigma(a^{\sharp}, y - a) = 0$ defines the unique support plane for N_a in a.
(c) $\omega(y)$ has only one maximum on this hyperplane. It is attained at a.

If $n = 1$ one calculates $\sigma(x, y) = \sigma \cdot x \cdot y$, $\sigma \neq 0$, $Y = \{y \in \mathbf{R}; \ y/\sigma > 0\}$, $y^\sharp = |y|/\sigma$ and $N_a = \{a\}$. Therefore the claims are obvious in this case.

We divide the proofs into several propositions and assume $n > 1$.

Proposition 1. *Each $y \in N_a$, $y \neq a$ satisfies $\sigma(a^\sharp, y - a) > 0$.*

Proof. By (3) we have for $\omega(y) = \omega(a)$, $y \neq a$

$$1 < e^{\sigma(a^\sharp, y - a)},$$

hence $\sigma(a^\sharp, y - a) > 0$. □

Proposition 2. *Given $x \in X$ then $\sigma(x, y - a) \geq 0$ for all $y \in N_a$ implies $x \in \overline{Y}$.*

Proof. For every $b \in Y$ the element $y = \sqrt[n]{\frac{\omega(a)}{\omega(b)}} b$ belongs to N_a, hence

$$\sqrt[n]{\frac{\omega(a)}{\omega(b)}} \sigma(x, b) \geq \sigma(x, a), \quad \text{or} \quad \sigma(x, b) \geq \sqrt[n]{\frac{\omega(b)}{\omega(a)}} \sigma(x, a).$$

As $b \to u \in \partial Y$ we have

$$\sigma(x, u) \geq 0 \quad \text{for} \quad u \in \partial Y$$

because of $\omega(u) = 0$. Since Y is a convex cone, for $y \in Y$ there exist two points $u_1, u_2 \in \partial Y$ such that $y = u_1 + u_2$ according to equation (11) in §2, consequently, $\sigma(x, y) \geq 0$ for $y \in Y$. Theorem 1 shows $x \in \overline{Y}$. □

Proposition 3. *$y \in Y$ implies $y^\sharp \in Y$.*

Proof. We substitute $x = a^\sharp$ in Proposition 2. Then we conclude from Proposition 1 that $a^\sharp \in \overline{Y}$ for $a \in Y$. Since $y \mapsto y^\sharp$ is an open mapping, the image of Y under $y \mapsto y^\sharp$ is open and consequently contained in $\overset{\circ}{\overline{Y}} = Y$. □

Proposition 4. *If $x \in X$, $x \neq 0$, $b \in Y$ then $\sigma(x, y - a) \geq 0$ for all $y \in N_a$ and $\sigma(x, b - a) = 0$ implies $\omega(b) \leq \omega(a)$.*

Proof. By Proposition 2 we have $x \in \overline{Y}$ and Theorem 1(a) yields

$$\sigma(x, a) > 0.$$

Since $y = \sqrt[n]{\frac{\omega(a)}{\omega(b)}} b$ belongs to N_a, it follows that

$$\sigma(x, a) = \sigma(x, b) = \sqrt[n]{\frac{\omega(b)}{\omega(a)}} \sigma(x, y) \geq \sqrt[n]{\frac{\omega(b)}{\omega(a)}} \sigma(x, a)$$

and $\omega(b) \leq \omega(a)$. □

Proof of Theorem 6. Let

$$H_x = \{y \in X; \sigma(x, y - a) = 0\}$$

be a support plane for N_a and $\mu(x) = \sup\{\omega(y); \ y \in H_x\}$. By Proposition 4 we have $\mu(x) = \omega(a)$. Let $u \in X$ and $\sigma(x, u) = 0$. For sufficiently small λ, the vector $b = a + \lambda u$ belongs to Y and H_x. By Proposition 4 we have $\omega(a + \lambda u) \le \omega(a)$ and from (2) it follows that

$$\log \omega(a) \ge \log \omega(a + \lambda u) = \log \omega(a) + \lambda \sigma(a^\sharp, u) + O(\lambda^2),$$

hence $0 \ge \lambda \sigma(a^\sharp, u) + O(\lambda^2)$. Dividing by $|\lambda|$ we get $\pm \sigma(a^\sharp, u) \le O(\lambda)$ and $\sigma(a^\sharp, u) = 0$. We have shown

$$u \in X, \quad \sigma(x, u) = 0 \quad \Rightarrow \quad \sigma(a^\sharp, u) = 0$$

and consequently $x = \lambda a^\sharp, \lambda \ne 0$. The hyperplane $H_{a^\sharp}$ is therefore the unique support plane in a and (b) has been proved.

By Proposition 1, $H_{a^\sharp} \cap N_a = \{a\}$ and $N_a \subset H_{a^\sharp}^+ := \{y; \sigma(a^\sharp, y - a) \ge 0\}$ hold. In view of $\sigma(a^\sharp, 0 - a) = -\sigma(a^\sharp, a) < 0$, we have $0 \notin H_{a^\sharp}^+$. This means that N_a is convex with respect to the origin. Thus (a) has been proved.

Let $y \in Y \cap H_{a^\sharp}$, $y \ne a$. In view of $\sqrt[n]{\frac{\omega(a)}{\omega(y)}} y \in N_a$ and (b) it follows that

$$\sigma\left(a^\sharp, \sqrt[n]{\frac{\omega(a)}{\omega(y)}} y - a\right) > 0, \quad \sigma(a^\sharp, y - a) = 0,$$

and therefore

$$\left(\sqrt[n]{\frac{\omega(a)}{\omega(y)}} - 1\right) \sigma(a^\sharp, a) > 0,$$

and this is (c).   □

Proposition 5. *The map $Y \to Y$, $y \mapsto y^\sharp$, is surjective.*

Proof. We consider $\omega(y)$ on the hyperplane $H = \{y \in X; \ \sigma(b, y) = 1\}$ for some $b \in Y$. By Lemma 1 we have for $y \in \overline{Y} \cap H$

$$1 = \sigma(b, y) \ge \varrho(b) \cdot |y|.$$

Therefore $H \cap \overline{Y}$ is compact. Let $a \in \overline{Y} \cap H$ satisfy

$$\omega(a) = \sup\{\omega(y); y \in H \cap \overline{Y}\}.$$

In view of $\omega(x) = 0$ for $x \in \partial Y$ it follows that $a \in Y$. Let $z \in N_a$, then one has

$$y = \frac{1}{\sigma(b, z)} z \in H \cap Y$$

and therefore

$$\omega(a) \geq \omega(y) = \frac{1}{\sigma^n(b, z)} \omega(z) = \frac{1}{\sigma^n(b, z)} \omega(a).$$

Consequently $\sigma(b, z) \geq 1 = \sigma(b, a)$ holds or

$$\sigma(b, z - a) \geq 0 \quad \text{for} \quad z \in N_a.$$

Hence $\sigma(b, z - a) = 0$ defines a support plane for N_a, and by Theorem 6(b) and (1) there is a $\lambda > 0$ such that

$$a^{\sharp} = \lambda b, \quad \text{or} \quad b = (\lambda a)^{\sharp}.$$

$\square$

Proposition 6. *The map $Y \to Y$, $y \to y^{\sharp}$, is injective.*

Proof. Let $a, b \in Y$ such that $a^{\sharp} = b^{\sharp}$. By (3) we have

$$\frac{\omega(b)}{\omega(a)} \leq e^{\sigma(a^{\sharp}, b-a)} = e^{-\sigma(b^{\sharp}, a-b)} \leq \frac{\omega(b)}{\omega(a)}.$$

Hence

$$\frac{\omega(b)}{\omega(a)} = e^{\sigma(a^{\sharp}, b-a)}$$

and therefore $a = b$ in view of (3). $\square$

Proof of Theorem 5. Combine Propositions 3, 5 and 6 as well as Lemma 5. $\square$

In view of $(\lambda y)^{\sharp} = \frac{1}{\lambda} y^{\sharp}$ due to (1), the Euler differential equation leads to

$$\frac{\partial y^{\sharp}}{\partial y} y = -y^{\sharp},$$

and it follows that

(4) $$y^{\sharp} = H(y)y, \quad y \in Y.$$

Now we conclude

$$\left(\frac{\partial}{\partial y}\log(\omega(y)\omega(y^\sharp))\right)^* = \left(\frac{\partial}{\partial y}\log\omega(y)\right)^* + \left(\frac{\partial}{\partial y}\log\omega(y^\sharp)\right)^*$$

$$= y^\sharp + \left(\frac{\partial y^\sharp}{\partial y}\right)^*\left(\frac{\partial}{\partial y^\sharp}\log\omega(y^\sharp)\right)^*$$

$$= y^\sharp - H(y)y^{\sharp\sharp},$$

because $H(y)$ is self-adjoint. (4) yields

(5) $$\left(\frac{\partial}{\partial y}\log(\omega(y)\omega(y^\sharp))\right)^* = H(y)(y - y^{\sharp\sharp}), \quad y \in Y.$$

Lemma 6. *The following four statements are equivalent:*

(A) $y^{\sharp\sharp} = y$ for all $y \in Y$.

(B) $\omega(y)\omega(y^\sharp)$ is constant on Y.

(C) $H(y)H(y^\sharp) = \mathrm{Id}$ for all $y \in Y$.

(D) The line element

$$ds = \sqrt{\sigma(H(y)\dot{y}, \dot{y})}dt$$

is invariant under the map $Y \to Y$, $y \mapsto y^\sharp$.

Proof. "(A) $\Leftrightarrow$ (B)": Use (5) and $\det H(y) \neq 0$.

"(C) $\Leftrightarrow$ (A)": (A) is equivalent to

$$\mathrm{Id} = \frac{\partial y}{\partial y} = \frac{\partial y^{\sharp\sharp}}{\partial y} = \frac{\partial y^{\sharp\sharp}}{\partial y^\sharp}\frac{\partial y^\sharp}{\partial y} = H(y^\sharp)H(y).$$

"(C) $\Leftrightarrow$ (D)": Let $y = y(\tau)$ be a curve in Y. Then one has

$$\frac{d}{dt}y^\sharp = \frac{\partial y^\sharp}{\partial y}\dot{y} = -H(y)\dot{y}$$

and consequently, (D) is equivalent to

$$H(y)H(y^\sharp)H(y) = H(y).$$

In view of $\det H(y) \neq 0$ this is equivalent to (C).   □

Theorem 7. *If the bilinear form σ is positive definite then the mapping $Y \to Y$, $y \mapsto y^\sharp$, has exactly one fixed point.*

Proof. We assume that

$$\sigma(x, x) > 0 \quad \text{for} \quad 0 \neq x \in X.$$

Then

$$|x| := \sqrt{\sigma(x, x)}$$

yields a norm $|x|$ on X, which is real-analytic for $x \neq 0$.

Now we consider the infimum of $|y|$ for $y \in N_a$. Since the norm ω vanishes on the boundary of Y, the set N_a is closed in X. Hence the distance from the origin measured by $|x|$ is positive and there exists $b \in N_a$ such that

$$|b| = \inf\{|y|; \ y \in N_a\}.$$

By elementary calculus, the tangent hyperplane H of N_a at b is orthogonal to b, thus $\sigma(b, y - b) = 0$ for all $y \in H$. On the other hand, $\sigma(b^\sharp, y - b) = 0$ for all $y \in H$ by Theorem 6(b), because H is the support plane at b.

This shows that b and $b^\sharp$ are linearly dependent elements of Y, hence $b^\sharp = \beta b$ for some $\beta > 0$. Now $c := \sqrt{\beta}b$ yields $c^\sharp = c$ due to (1).

Concerning uniqueness, let $v \in Y$ such that $v^\sharp = \lambda v$ for some $\lambda > 0$. The support plane of N_v at v is given by the equation $\sigma(v, y - v) = 0$.

Let $z \in N_v$, $z \neq v$. Then the origin and z lie on different sides of the support plane by Theorem 6(b). This shows $|z| > |v|$, hence $|\cdot|$ attains its minimum on N_v at v. Homogeneity now shows that v is a multiple of b.   □

§9. Homogeneous domains of positivity

A domain of positivity Y with respect to σ is called *homogeneous* if for any two points $a, b \in Y$, there exists $W \in \Sigma(Y)$ such that

$$b = Wa,$$

i.e. if the group $\Sigma(Y)$ acts transitively on Y. We now assume that Y is homogeneous. Consider a function $\varphi(y)$ such that $\varphi(Wy) = |\det W| \cdot \varphi(y)$ for $W \in \Sigma(Y)$. The quotient $\varphi(y)/\omega(y)$, with ω a special norm on Y from Theorem 4, is invariant under the maps $y \mapsto Wy$, $W \in \Sigma(Y)$, and consequently the quotient is constant. We get $\varphi(y) = \gamma \cdot \omega(y)$.

In particular, we have: *Two norms on a homogeneous domain of positivity only differ by a constant factor.*

Now we consider

$$\varphi(y) := \frac{1}{\omega(y^\sharp)}.$$

From Lemma 5 we have for $W \in \Sigma(Y)$

$$\varphi(Wy) = \frac{1}{\omega((Wy)^\sharp)} = \frac{1}{\omega(W^{*-1}y^\sharp)} = \frac{|\det W|}{\omega(y^\sharp)} = |\det W| \cdot \varphi(y).$$

Consequently,

$$\omega(y) \cdot \omega(y^\sharp) = \text{constant}$$

and from Lemma 6 we obtain

Theorem 8. *Let Y be a homogeneous domain of positivity, then:*

(a) $\omega(y) \cdot \omega(y^\sharp)$ is constant on Y.

(b) $y^{\sharp\sharp} = y$ for all $y \in Y$.

(c) $H(y)H(y^\sharp) = \text{Id}$ for all $y \in Y$.

(d) The line element ds is invariant under the map $Y \to Y$, $y \mapsto y^\sharp$.

The definition of the special norm $\omega(y)$ in Theorem 4 involves a factor ω_0, which can be chosen arbitrarily. For homogeneous domains of positivity we fix our choice of ω_0. The definitions of $y^\sharp$ and $H(y)$ do not depend on ω_0. In view of

$$\omega(y) \cdot \omega(y^\sharp) = \omega_0^2/[\tilde{\omega}(y) \cdot \tilde{\omega}(y^\sharp)] = \omega_0^2 \cdot \text{constant},$$

we determine ω_0 in such a way that

$$\omega(y) \cdot \omega(y^\sharp) = 1 \quad \text{for all} \quad y \in Y.$$

Theorem 9. *Let Y be a homogeneous domain of positivity with respect to σ and let σ be positive definite. Then:*

(a) For the fixed point c of the involution $Y \to Y$, $y \mapsto y^\sharp$, we have $H(c) = \text{Id}$.

(b) $y \in Y$ implies $H(y) \in \Sigma(Y)$.

(c) $u > v > 0$ implies $v^\sharp > u^\sharp > 0$.

Proof. (a) By Theorem 7, the involution $Y \to Y, y \mapsto y^\sharp$, has a unique fixed point c. From Theorem 8(c) we therefore get $H^2(c) = \text{Id}$. By equation (4) in §7 the linear transformation $H(c)$ is positive definite and by assumption the bilinear form σ is positive definite. Therefore one knows from Linear Algebra that $H(c)$ is the identity.

(b) In equation (5) in §7 we proved that

$$W^* H(Wy)W = H(y) \quad \text{for} \quad W \in \Sigma(Y) \quad \text{and} \quad y \in Y.$$

Since $\Sigma(Y)$ acts transitively on Y, for fixed $y \in Y$ we find a transformation $W \in \Sigma(Y)$ such that $Wy = c$, hence $H(y) = W^*W$. By Lemma 2 we have $W^* \in \Sigma(Y)$ and therefore $H(y) \in \Sigma(Y)$.

(c) Since $y \mapsto y^\sharp$ is real-analytic on Y, we get for small $x \in X$ and $a \in Y$

$$(a+x)^\sharp = a^\sharp + \frac{\partial y^\sharp}{\partial y}\bigg|_{y=a} x + O(|x|^2)$$

or

(1)
$$(a+x)^\sharp = a^\sharp - H(a)x + O(|x|^2).$$

Proposition 7. *Given $a, x \in Y$ there exists an $\varepsilon > 0$ such that for $\tau_2 > \tau_1$ and $|\tau_\nu| < \varepsilon$ we have*

$$(a+\tau_1 x)^\sharp > (a+\tau_2 x)^\sharp.$$

Proof. From (1) we conclude for sufficiently small $\varepsilon > 0$

$$(a+\tau_\nu x)^\sharp = a^\sharp - \tau_\nu H(a)x + O(\varepsilon^2)$$

and therefore

$$(a+\tau_1 x)^\sharp - (a+\tau_2 x)^\sharp = (\tau_2 - \tau_1)H(a)x + O(\varepsilon^2).$$

In view of $H(a) \in \Sigma(Y)$ we have

$$(a+\tau_1 x)^\sharp - (a+\tau_2 x)^\sharp \in Y \quad \text{for} \quad \tau_2 > \tau_1 \quad \text{and} \quad |\tau_\nu| < \varepsilon.$$

This, indeed, gives the proposition. □

Now let $u > v > 0$. The curve

$$y(\tau) = v + \tau(u - v), \quad 0 \leq \tau \leq 1,$$

in Y can be covered by the y–images of a finite number of (relatively) open subsets U_ν of the interval $[0; 1]$ such that

$$y^\sharp(\tau_1) > y^\sharp(\tau_2) \quad \text{for} \quad \tau_2 > \tau_1 \quad \text{and} \quad \tau_1, \tau_2 \in U_\nu.$$

Consequently, $v^\sharp = y^\sharp(0) > y^\sharp(1) = u$ follows. □

Notes

The notion of a domain of positivity was introduced in 1957 by the author [16] and the basic notions were given in this paper. The present lecture notes follow a manuscript of E. Artin, H. Braun and M. Koecher [0]. The emphasis is put on getting as many results as possible in the general case of not necessarily homogeneous domains of positivity. O.S. Rothaus [22] and E.B. Vinberg [26] studied homogeneous domains of positivity from the point of view of homogeneous spaces and Lie groups.

Editors' Notes

1. (i) The "Schwarz inequality" sometimes referred to is the inequality (for suitable κ)

$$|\sigma(x,y)| \leq \kappa \cdot |x| \cdot |y| \quad \text{for all} \quad x \quad \text{and} \quad y.$$

It is a consequence of the compactness of the unit sphere.

(ii) Since $\Sigma(Y) \subseteq GL(X)$ is closed, it is actually a Lie subgroup of $GL(X)$.

(iii) Considering (3) in §5 we remark that a $\psi : GL(X) \to K\backslash\{0\}$ is a group homomorphism if and only if a group homomorphism $\partial : K\backslash\{0\} \to K\backslash\{0\}$ exists such that $\psi(A) = \partial(\det A)$, $A \in GL(X)$ (cf. [Ko3], 3.4.8).

(iv) Relation (1) in §7 is also an easy consequence of (2), via differentiation.

2. In the language used today, a domain of positivity is a self-dual open proper convex cone. Generally, the dual of an open convex cone Y is the set $Y^* := \{\ell \in X^*; \ell(y) > 0 \text{ for all } y \in Y\}$. Here, X^* is identified with X through the form σ. Vinberg [Vi] proved that most of the elementary properties also hold for open proper convex cones. He also noted that for every such cone Z the cone $Z \times Z^* \subset X \times X^*$ is a domain of positivity. It seems, though, that domains of positivity with respect to a positive definite bilinear form are the most interesting ones. A recent introduction to these homogeneous domains of positivity is contained in Faraut and Korányi [FKo]. Vinberg [Vi] constructed and, to some extent, classified arbitrary homogeneous open convex proper cones. His starting point was the observation that the normalizer of the identity component of the automorphism group of such a cone is a real algebraic group (with the same identity component). This allowed him to use his theorem on a "polar decomposition" (compact triangular) of real algebraic groups to establish a correspondence between the cones in question and certain left-symmetric algebras ("clans"), which

in turn made classification possible. Later, Dorfmeister [Do1], [Do3], [Do4] succeeded in giving a precise algebraic description and classification of these cones, which again involved Jordan algebras in an essential manner. See also the article [DK2] by Dorfmeister and Koecher. Concerning applications in analysis, Gindikin's article [Gi1] should be mentioned. A collection of classical and recent results is contained in [FKo]. In the past few years, there has been much work on applications of domains of positivity to statistics. This started even before the introduction of Jordan algebras and domains of positivity with Wishart's classical work [Wi] on the cone of positive definite symmetric matrices. Among recent contributions we mention Bernadac [Be], Casalis [Cas], Casalis and Letac [CL], Letac and Massam [LM], Massam and Neher [MNe1], [MNe2], and Neher [Ne1], [Ne2].

Chapter II. Omega Domains

§1. The notion of an ω–domain.

In this chapter, let X be a vector space over $\mathbf{R}$ of finite dimension n and let $|x|$ be a norm on X. Generalizing the notion of a domain of positivity, we consider a pair (Y, ω), which has the following properties:

(D.1) $Y \neq \emptyset$ is an open and connected subset of X satisfying $\lambda y \in Y$ for every $\lambda > 0$ and every $y \in Y$.

(D.2) $\omega : \overline{Y} \to \mathbf{R}$ is a function such that:

(a) $\omega(y)$ is a real-analytic and positive on Y.

(b) $\omega(y)$ is continuous on $\overline{Y}$ and vanishes on the boundary ∂Y.

(c) $\omega(\lambda y) = \lambda^n \omega(y)$ for $\lambda > 0$ and $y \in Y$.

(d)   The bilinear form $\Delta_y^u \Delta_y^v \log \omega(y)$ on X is non-singular for every $y \in Y$.

We fix a point $c \in Y$ and put

$$\sigma(u, v) = -\Delta_y^u \Delta_y^v \log \omega(y)\big|_{y=c}.$$

Since σ remains unchanged if we replace $\omega(y)$ by $\gamma \cdot \omega(y)$, $\gamma > 0$, we may assume that $\omega(c) = 1$. The symmetric bilinear form $\sigma(u, v)$ on X is non-singular due to (D.2). Since $\Delta_y^u \log \omega(y)$ is a linear form on X, there exists for each $y \in Y$ a point $y^\sharp \in X$ such that

$$\sigma(u, y^\sharp) = \Delta_y^u \log \omega(y) \quad \text{for all} \quad u \in X.$$

$y^\sharp$ depends real-analytically on y. On the other hand, $\Delta_y^u \Delta_y^v \log \omega(y)$ is a symmetric non-singular bilinear form for each $y \in Y$. Therefore there exists a self-adjoint (with respect to σ) linear transformation $H(y)$ such that

$$\sigma(H(y)u, v) = -\Delta_y^u \Delta_y^v \log \omega(y).$$

The transformation depends real-analytically on y, and satisfies $\det H(y) \neq 0$. The definition of $\sigma(u, v)$ shows that

$$H(c) = \text{Id}.$$

Using the definition of $y^\sharp$ we get

(1)
$$y^\sharp = \left(\frac{\partial \log \omega(y)}{\partial y} \right)^*, \quad H(y) = -\frac{\partial y^\sharp}{\partial y}.$$

Here * is to be taken with respect to the bilinear form σ.

Let $\Sigma = \Sigma(Y, \omega)$ be the group of all linear transformations $W : X \to X$ such that

$$WY = Y, \quad \omega(Wy) = |\det W| \cdot \omega(y) \quad \text{for} \quad y \in Y.$$

Since Y is open, the transformations $W \in \Sigma$ are non-singular. In view of (D.2) the mappings $y \mapsto \lambda y$, $\lambda > 0$, belong to Σ.

Lemma 1. *For all $W \in \Sigma$ and $y \in Y$ we have*

$$(Wy)^\sharp = W^{*-1} y^\sharp \quad \text{and} \quad W^* H(Wy) W = H(y).$$

Proof. We have

$$\sigma(u, y^\sharp) = \Delta_y^u \log \omega(y) = \Delta_{Wy}^{Wu} \log \omega(Wy)$$
$$= \sigma(Wu, (Wy)^\sharp) = \sigma(u, W^*(Wy)^\sharp)$$

and

$$\sigma(H(y)u, v) = -\Delta_y^u \Delta_y^v \log \omega(y) = -\Delta_y^u \Delta_y^v \log \omega(Wy)$$
$$= -\Delta_{Wy}^{Wu} \Delta_{Wy}^{Wu} \log \omega(Wy) = \sigma(H(Wy)Wu, Wv)$$
$$= \sigma(W^* H(Wy)Wu, v).$$

$\square$

In particular, we get $(\lambda y)^\sharp = \frac{1}{\lambda} y^\sharp$ for $\lambda > 0$. The Euler differential equation for this homogeneous function leads to $\frac{\partial y^\sharp}{\partial y} y = -y^\sharp$, hence

(2)
$$y^\sharp = H(y)y.$$

Thus $H(c) = \text{Id}$ implies $c^\sharp = c$.

Definition. Y is called an ω–*domain* if Y satisfies the axioms (D.1), (D.2) and

(D.3) For every $y \in Y$ the map $H(y)$ belongs to Σ.

Remark. Every result of this chapter remains valid if we drop the requirement that $\omega(y)$ vanishes on the boundary of Y.

In order to define the transformation $H(y)$ we need the bilinear form $\sigma(x, y)$ and therefore the arbitrary point $c \in Y$. If we want to indicate the choice of the point c, we will write (Y, ω, c) for the ω-domain; the bilinear form $\sigma(x, y)$ is said to be associated with the domain (Y, ω, c).

Let (Y, ω, c) be an ω-domain in X. Since the transformation $H(y)$ belongs to Σ and since (2) holds, it follows that $y^\sharp \in Y$ for every $y \in Y$. Putting $W = H(y) = W^*$ in Lemma 1 we get

$$(3) \qquad\qquad H(y^\sharp)H(y) = \mathrm{Id}.$$

Applying this transformation to y and using (2) we obtain

$$(4) \qquad\qquad y^{\sharp\sharp} = y \quad \text{for} \quad y \in Y.$$

Therefore the mapping $Y \to Y$, $y \mapsto y^\sharp$, is bijective.

Lemma 2. *If Y is an ω-domain then $W^* \in \Sigma$ holds for all $W \in \Sigma$.*

Proof. Lemma 1 leads to $W^*H(Wy)W = H(y)$, with $W, H(y)$ and $H(Wy)$ belonging to Σ. $\qquad\qquad\qquad\qquad\qquad\qquad\qquad\qquad\qquad\qquad$ □

In view of

$$\left(\frac{\partial}{\partial y} \log\left(\omega(y)\,\omega(y^\sharp)\right) \right)^* = y^\sharp - H(y)y^{\sharp\sharp} = 0,$$

we conclude that $\omega(y)\omega(y^\sharp)$ is constant. Then $\omega(c) = \omega(c^\sharp) = 1$ gives

$$(5) \qquad\qquad \omega(y)\omega(y^\sharp) = 1.$$

Next, (2) and $H(y) \in \Sigma(Y, \omega)$ lead to $\omega(y^\sharp) = \omega(H(y)y) = |\det H(y)| \cdot \omega(y)$, hence

$$(6) \qquad\qquad |\det H(y)| = \frac{1}{\omega^2(y)}.$$

Theorem 1. *Let (Y, ω, c) be a domain in a vector space X satisfying (D.1) and (D.2). If the subgroup of $\Sigma(Y, \omega)$ consisting of all linear transformations, for which the adjoint transformation (with respect to σ) also belongs to $\Sigma(Y, \omega)$, acts transitively on Y, then (Y, ω, c) is an ω-domain.*

Proof. Given $y \in Y$ we find $W \in \Sigma(Y, \omega)$ such that $Wy = c$, and $W^* \in \Sigma(Y, \omega)$. Applying Lemma 1 we get $H(y) = W^*W \in \Sigma(Y, \omega)$. $\qquad\qquad$ □

Another class of ω–domains was described in Theorem I.9:

Theorem 2. *Let Y be a homogeneous domain of positivity with respect to the positive definite bilinear form $\sigma(x, y)$, let $\omega(y)$ be a special norm on Y and let c be the unique fixed point of the involution $y \mapsto y^\sharp$ on Y. Then we have:*

(a) (Y, ω, c) is an ω–domain and the group $\Sigma(Y, \omega)$ coincides with $\Sigma(Y)$.

(b) The involution $y \mapsto y^\sharp$ of the domain of positivity is equal to the involution of the ω–domain.

(c) The given bilinear form $\sigma(x, y)$ is equal to the bilinear form associated with the ω–domain (Y, ω, c).

§2. Some examples

(A) Let $\mu(x, y)$ be a symmetric non-singular bilinear form on X. We consider the set $\{y \in X;\ \mu(y, y) > 0\}$, and assume that this set is non-empty. Let Y denote a connected component of this set. Putting

$$\omega(y) = [\mu(y, y)]^{n/2}$$

we get a pair (Y, ω) satisfying the properties (D.1) and (D.2). To verify this, we only need to prove that the bilinear form $\Delta_y^u \Delta_y^u \log \omega(y)$ is non-singular. Let us compute this bilinear form:

$$\Delta_y^u \log \omega(y) = \frac{n}{2} \cdot \Delta_y^u \log \mu(y, y) = n \cdot \frac{\mu(u, y)}{\mu(y, y)},$$

$$\Delta_y^v \Delta_y^u \log \omega(y) = n \cdot \Delta_y^v \frac{\mu(u, y)}{\mu(y, y)}$$

$$= \frac{n}{\mu^2(y, y)} [\mu(y, y) \cdot \mu(u, v) - 2\mu(u, y) \cdot \mu(v, y)].$$

If this bilinear form is singular for some $y \in Y$, there exists $u \neq 0$ such that $\mu(y, y) \cdot \mu(u, v) = 2\mu(u, y) \cdot \mu(v, y)$ for all $v \in X$. Setting $v = y$ we obtain $\mu(u, y) = 0$ from $\mu(y, y) > 0$. This yields $\mu(u, v) = 0$ for all $v \in X$ as a contradiction. Therefore the bilinear form is non-singular.

Now let c be a point in Y such that $\mu(c, c) = 1$, and thus

$$\sigma(u, v) = -\Delta_y^u \Delta_y^v \log \omega(y)\big|_{y=c} = n [2\mu(u, c) \cdot \mu(v, c) - \mu(u, v)].$$

In order to compute the transformation $H(y)$, we start with $-\Delta_y^u \Delta_y^v \log \omega(y) = \sigma(H(y)u, v)$ and obtain

$$\frac{1}{\mu^2(y,y)} [2\mu(u,y) \cdot \mu(v,y) - \mu(y,y) \cdot \mu(u,v)]$$
$$= 2\mu(H(y)u,c) \cdot \mu(v,c) - \mu(H(y)u,v).$$

This equation holds for all $v \in X$ and consequently

(1) $\qquad \frac{1}{\mu^2(y,y)} [2\mu(u,y)y - \mu(y,y)u] = 2\mu(H(y)u,c)c - H(y)u.$

Apply $\mu(\cdot,c)$ to this equation. The result is

$$\mu(H(y)u,c) = \frac{1}{\mu^2(y,y)} [2\mu(u,y) \cdot \mu(y,c) - \mu(y,y) \cdot \mu(u,c)].$$

Substituting this in (1) we obtain

(2) $\qquad H(y)u = \frac{1}{\mu^2(y,y)} \{\mu(y,y)u - 2\mu(u,y)y +$

$$2[2\mu(u,y) \cdot \mu(y,c) - \mu(y,y) \cdot \mu(u,c)]c\}.$$

An elementary calculation leads to

(3) $\qquad \mu(H(y)u, H(y)u) = \frac{\mu(u,u)}{\mu^2(y,y)}, \quad |\det H(y)| = \frac{1}{\mu^n(y,y)} = \frac{1}{\omega^2(y)}.$

By virtue of $y^\sharp = H(y)y$ we get for $u = y$

(4) $\qquad y^\sharp = \frac{1}{\mu(y,y)} [-y + 2\mu(y,c)c], \quad \mu(y^\sharp, y^\sharp) = \frac{1}{\mu(y,y)}.$

Since Y is connected, $c^\sharp = c$, and $\mu(y^\sharp, y^\sharp) > 0$ for $y \in Y$, it follows that $y^\sharp \in Y$ for all $y \in Y$. Given $u \in Y$ we consider the points $H(y)u$ and $H^{-1}(y)u$. It follows from (3) that $\mu(H(y)u, H(y)u) > 0$ for $u \in Y$. Since there exists a curve in Y joining u and y and since $H(y)y = y^\sharp$ belongs to Y, we get $H(y)u \in Y$ or $H(y)Y \subset Y$. Analogous arguments yield $H^{-1}(y)Y \subset Y$, and thus $H(y)Y = Y$. Therefore $H(y)$ belongs to Σ, and Y is an ω–domain.

Taking $X = \mathbf{R}^n$, $Y = X \backslash \{0\}$, $\mu(x,y) = x^t y$ (the inner product), we get an ω–domain Y whenever $n > 1$.

(B) Let X be the set of all real symmetric $m \times m$ matrices x. Then X is a vector space over $\mathbf{R}$ of dimension $n = m(m+1)/2$. For every $x \in X$ the signature of x is defined to be (p,q), where p resp. q is the number of positive resp. negative eigenvalues of x. We set sign $x := (p,q)$. Let c be a fixed matrix in X satisfying $\det c \neq 0$ and $c^{-1} = c$, and let

$$Y = \{y \in X; \text{ sign } y = \text{ sign } c\}, \quad \omega(y) = |\det y|^{(m+1)/2}.$$

It is well-known that each $y \in Y$ can be written as $y = rcr^t$ with some real square matrix r and $\det r > 0$. Since the set of all real $m \times m$ matrices r

fulfilling $\det r > 0$ is open and connected, the set Y, being its image under the mapping $r \mapsto rcr^t$, is also open and connected. Concerning the properties (D.1) and (D.2) we only need to show that the form $\Delta_y^u \Delta_y^v \log \omega(y)$ is non-singular. With e denoting the unit matrix, we have

$$
\begin{aligned}
\Delta_y^u \log \omega(y) &= \frac{m+1}{2} \frac{d}{d\tau} \log \det(y + \tau u)\Big|_{\tau=0} \\
&= \frac{m+1}{2} \frac{d}{d\tau} \log \det(e + \tau y^{-1} u)\Big|_{\tau=0} = \frac{m+1}{2} \cdot \operatorname{tr}(y^{-1} u).
\end{aligned}
$$

Next

$$
\begin{aligned}
(y + \tau v)^{-1} &= y^{-1}(e + \tau v y^{-1})^{-1} \\
&= y^{-1}\left[e - \tau v y^{-1} + O(\tau^2)\right] = y^{-1} - \tau y^{-1} v y^{-1} + O(\tau^2)
\end{aligned}
$$

implies that

$$
\Delta_y^u \Delta_y^v \log \omega(y) = -\frac{m+1}{2} \cdot \operatorname{tr}(y^{-1} v y^{-1} u).
$$

This bilinear form is non-singular. Therefore Y satisfies (D.1) and (D.2). We put

$$
\sigma(u, v) = -\Delta_y^u \Delta_y^v \log \omega(y)\big|_{y=c} = -\frac{m+1}{2} \cdot \operatorname{tr}(cv \, cu)
$$

and get $\operatorname{tr}(v y^{-1} u y^{-1}) = \operatorname{tr}(y^{-1} v y^{-1} u) = \operatorname{tr}(cvcH(y)u) = \operatorname{tr}(vcH(y)uc)$ or $cH(y)uc = y^{-1} u y^{-1}$, or (using formula (4) in Chapter I, §5)

$$
H(y)u = cy^{-1} u y^{-1} c, \quad |\det H(y)| = |\det y|^{-(m+1)}.
$$

Thus we have $H(y)Y = Y$. In view of $|\det H(y)v| = |\det v|/(\det y)^2$ the transformation $H(y)$ belongs to Σ, and Y is an ω–domain.

Since the linear transformation $y \mapsto W(r)y = ryr^t$ belongs to Σ for any $m \times m$ matrix r with $\det r \neq 0$, it follows that Σ acts transitively on Y.

§3. The geodesics of an ω–domain

We need some notions from the calculus of variations. Let Y be an open subset of the vector space X and let $\lambda(y, x)$ be a continuously differentiable real-valued function on $Y \times X$. A twice continuously differentiable curve $y = y(\tau)$, $\tau_1 \leq \tau \leq \tau_2$, in Y is called a solution of the variational problem relative to the integral

$$
\int_{\tau_1}^{\tau_2} \lambda(y, \dot{y}) d\tau,
$$

if for every continuously differentiable mapping

$$y = y(\tau, \varepsilon), \quad \tau_1 \le \tau \le \tau_2, \quad 0 \le \varepsilon \le \varepsilon_1,$$
$$y(\tau, 0) = y(\tau), \quad y(\tau_1, \varepsilon) = y(\tau_1), \quad y(\tau_2, \varepsilon) = y(\tau_2),$$

the function

(1)
$$\rho(\varepsilon) = \int_{\tau_1}^{\tau_2} \lambda \left(y(\tau, \varepsilon), \frac{\partial y(\tau, \varepsilon)}{\partial \tau} \right) d\tau$$

fulfills the condition

$$\dot\rho(0) = 0.$$

This variational problem gives rise to the Euler–Lagrange differential equation

$$\frac{\partial \lambda}{\partial y} = \frac{d}{d\tau} \frac{\partial \lambda}{\partial \dot y}, \quad \lambda = \lambda(y, \dot y).$$

It is well-known that the curve $y = y(\tau)$ is a solution of the variational problem if and only if the Euler–Lagrange differential equations are satisfied.

Now let (Y, ω) be an ω-domain in X and let

$$\lambda(y, x) = \sigma(H(y)x, x).$$

Each solution of the variational problem relative to λ is called a *geodesic of* Y. The geodesics of Y are therefore exactly the solutions of the differential equations

$$\frac{d}{d\tau} \frac{\partial \lambda(y, \dot y)}{\partial \dot y} = \frac{\partial \lambda(y, \dot y)}{\partial y}.$$

By the definition of $\lambda(y, \dot y)$ we have

$$2 \frac{d}{d\tau} H(y) \dot y = \left(\frac{\partial \sigma(H(y) \dot y, \dot y)}{\partial y} \right)^*,$$

which leads to

(2)
$$\ddot y = k(y, \dot y),$$

where

$$k : Y \times X \to X$$

is a function which is real-analytic in the first argument and a polynomial of degree 2 in the second argument.

If we change the parameter τ to $\alpha\tau+\beta$, $\alpha \neq 0$, the integral (1) is multiplied by a factor α. Hence we have: If $y = y(\tau)$ is a geodesic and α, $\beta \in \mathbf{R}$, $\alpha \neq 0$, then $y(\alpha\tau + \beta)$ is a geodesic, too. It is known from Lemma 1 that $W^*H(Wy)W = H(y)$ for $W \in \Sigma$, and therefore we get $\lambda(Wy, Wx) = \lambda(y, x)$. It follows that $Wy(\tau)$ is a geodesic provided that $W \in \Sigma$ and $y(\tau)$ is a geodesic. We now prove that $y^\sharp(\tau)$ is a geodesic provided that $y(\tau)$ is a geodesic. Indeed, let us consider the curve $y^\sharp(\tau)$. By virtue of $\hat{y^\sharp} = \frac{\partial y^\sharp}{\partial y}\dot{y} = -H(y)\dot{y}$, we get

$$\lambda(y^\sharp, \hat{\dot{y^\sharp}}) = \sigma(H(y^\sharp)H(y)\dot{y}, H(y)\dot{y}) = \sigma(\dot{y}, H(y)\dot{y}) = \lambda(y, \dot{y}).$$

Lemma 3. *Given $a \in Y$ there exists a mapping $x \mapsto g_a(x)$ of a neighborhood V of the origin in X onto a neighborhood U of a in Y such that*

(a) $g_a(0) = a$, $\frac{\partial g_a(x)}{\partial x}\big|_{x=0} = \mathrm{Id}$.

(b) $x \mapsto g_a(x)$ is bijective and real-analytic.

(c) In U, each geodesic $y(\tau)$ through the point a can be represented as

$$y(\tau) = g_a(\tau u)$$

for a suitable $u \in X$.

Proof. By (2), the differential equations for the geodesics can be written in the form

$$\ddot{y} = k(y, \dot{y}),$$

where $k(y, x)$ is real-analytic in y and a polynomial of degree 2 in x.

The existence and uniqueness theorems for differential equations show that for given initial values

$$y(0) = a, \quad \dot{y}(0) = u, \quad a \in Y, \quad u \in X,$$

there exists exactly one geodesic. For all (a, u) in a suitable neighborhood of a point in $Y \times X$, the solutions are real-analytic in τ, a and u, provided that τ belongs to a certain neighborhood of the origin.

We fix $a \in Y$ and consider a suitable neighborhood V' of 0 in X. The unique geodesic $y(\tau)$ satisfying $y(0) = a$, $\dot{y}(0) = u$ with $u \in V'$ will be denoted by $g_a(\tau, u)$, thus

$$g_a(0, u) = a, \quad \dot{g}_a(0, u) = u.$$

Since $g_a(\alpha\tau, u)$, $0 < |\alpha| \leq 1$, is also a geodesic, with the initial values a and αu, it follows that

(3) $$g_a(\alpha\tau, u) = g_a(\tau, \alpha u), \quad |\alpha| \le 1.$$

Now we consider the expansion of $g_a(\tau, u)$ as a function of τ:

$$g_a(\tau, u) = \sum_{\nu=0}^{\infty} \frac{1}{\nu!} b_\nu(u) \tau^\nu.$$

Since $g_a(\tau, u)$ is real-analytic in u, the coefficients $b_\nu(u)$ are real-analytic functions on V'. Using (3) we get $b_\nu(\alpha u) = \alpha^\nu b_\nu(u)$. Therefore $b_\nu(u)$ is a homogeneous polynomial of degree ν in the components of u. Thus we have

$$g_a(\tau, u) = g_a(\tau u),$$

if we put

$$g_a(x) := \sum_{\nu=0}^{\infty} \frac{1}{\nu!} b_\nu(x).$$

This expansion is absolutely convergent in a neighborhood $V'' \subset V'$ of 0 due to the convergence properties of $g_a(\cdot, \cdot)$. In view of $g_a(0, u) = a$ and $\dot{g}_a(0, u) = u$ we get

$$b_0(x) = a, \quad b_1(x) = x.$$

But this is claim (a) and by construction we also get (c). In view of

$$\left. \frac{\partial g_a(x)}{\partial x} \right|_{x=0} = \text{Id}$$

we find a neighborhood $V \subset V''$ of 0, such that the mapping $x \mapsto g_a(x)$ maps V bijectively onto a neighborhood U of a in Y. This completes the proof. $\square$

Let $y(\tau)$, $0 \le \tau \le 2\alpha$, be a geodesic. We consider the curves

$$y_1(\tau) = y^\sharp(\tau), \quad y_2(\tau) = H(y(\alpha)) y(2\alpha - \tau)$$

in the interval $0 \le \tau \le 2\alpha$. Both curves are geodesics. (Recall that $H(y(\alpha)) \in \Sigma$.) By virtue of

$$y_1(\alpha) = y^\sharp(\alpha) = H(y(\alpha))y(\alpha) = y_2(\alpha),$$
$$\dot{y}_1(\alpha) = -H(y(\alpha))\dot{y}(\alpha) = \dot{y}_2(\alpha),$$

the curves coincide:

(4) $$y^\sharp(\tau) = H(y(\alpha))y(2\alpha - \tau), \quad 0 \le \tau \le 2\alpha.$$

Lemma 3 implies that for given $a \in Y$ the map $u \mapsto g_a(u)$ from a neighborhood V of 0 onto a neighborhood U of a in Y is real-analytic on V and that $g_a(\tau u)$ is a geodesic provided that $\tau u \in \overline{V}$.

Substituting $y(\tau) = g_a(\tau u), \alpha = \frac{1}{2}$, in (4) we get for $u \in V$ and $\tau = 1$

(5) $$g_a^\sharp(u) = H\big(g_a\big(\tfrac{1}{2}u\big)\big)g_a(0) = H\big(g_a\big(\tfrac{1}{2}u\big)\big)a.$$

We consider the neighborhood

$$V' = \{u; \tfrac{1}{2}u \in V\}$$

and put for $u \in V'$

$$h(u) = \big(H\big(g_a\big(\tfrac{1}{2}u\big)\big)a\big)^\sharp.$$

Since $g_a(u)$ and $H(y)$ are real-analytic, $h : V' \to Y$ is a real-analytic function. Given $u \in V$ it follows from (5) and equation (4) in §1 that

$$h(u) = \big(H\big(g_a\big(\tfrac{1}{2}u\big)\big)a\big)^\sharp = g_a(u).$$

Therefore $h(u)$ is a real-analytic continuation of $g_a(u)$ into the neighborhood V'. Hence $h(\tau u)$ is also a geodesic. Repeating this process we see that $g_a(u)$ can be extended to a real-analytic function on X. This continuation will again be denoted by $g_a(u)$. Therefore we get a real-analytic mapping

$$g_a : X \to Y$$

such that $g_a(\tau u)$ is geodesic for every u. Every geodesic through a can be written in the form $g_a(\tau u)$. We apply (4) to the geodesic $g_a(\tau u)$ and obtain for all τ and α in $\mathbf{R}$:

$$g_a^\sharp(\tau u) = H(g_a(\alpha u))g_a((2\alpha - \tau)u).$$

Substituting $\tau = -\xi - \eta$, $\alpha = -\frac{\xi}{2}$, we obtain

$$g_a^\sharp(-(\xi + \eta)u) = H(g_a(-\tfrac{\xi}{2}u))g_a(\eta u).$$

Especially for $\xi = 0$ and $\eta = -1$ we have

(6) $$g_a^\sharp(u) = H(a)g_a(-u).$$

Substituting this into the formula above we get

(7) $$H(a)g_a((\xi + \eta)u) = H(g_a(-\tfrac{\xi}{2}u))g_a(\eta u).$$

Putting $a = c$, $g(u) = g_c(u)$ and summarizing we obtain

Theorem 3. *Let (Y, ω, c) be an ω-domain. Then there exists a real-analytic mapping $X \to Y, u \mapsto g(u)$, such that the following statements hold:*

(a) $g(0) = c$, $\frac{\partial g(u)}{\partial u}\big|_{u=0} = \mathrm{Id}$.

(b) $g^\sharp(u) = g(-u)$.

(c) $g((\xi + \eta)u) = H(g(-\tfrac{\xi}{2}u))g(\eta u)$ for every $\xi, \eta \in \mathbf{R}$.

(d) The curves $g(\tau u)$ are exactly the geodesics through c.

§4. Non-associative algebras

In the next section we will prove that there is a relation between ω–domains and certain non-associative algebras. For this reason we introduce some notions about algebras.

In the whole section let X be a finite-dimensional vector space over a field K. Moreover let $X \times X \to X, (u, v) \mapsto u \circ v$, be a mapping. The pair $A = (X, \circ)$ is called an *algebra* (over K) if

$$(u_1 + u_2) \circ v = (u_1 \circ v) + (u_2 \circ v), \ u \circ (v_1 + v_2) = (u \circ v_1) + (u \circ v_2),$$
$$(\alpha u) \circ v = \alpha(u \circ v) = u \circ (\alpha v), \ \alpha \in K.$$

We also write $u \in A$ for $u \in X$. Given $u \in A$ the mapping $X \to X, v \mapsto u \circ v$, is a linear transformation $L(u)$ of X, in formulas

$$(1) \qquad\qquad L(u)v = u \circ v,$$

and we get

$$(2) \qquad L(u_1 + u_2) = L(u_1) + L(u_2), \ L(\alpha u) = \alpha L(u), \quad \alpha \in K.$$

Vice versa, if we start with a linear transformation $L(u)$ of the vector space X fulfilling (2) and define a composition $u \circ v$ by (1), we obtain an algebra $A = (X, \circ)$.

The algebra A is called *commutative* if $u \circ v = v \circ u$ for every $u, v \in A$. This is equivalent to $L(u)v = L(v)u$.

The algebra A is called an *algebra with unit element* if there exists an element $c \in A$ such that

$$u \circ c = c \circ u = u$$

for every $u \in A$. It is clear that A has at most one unit element. In terms of the linear transformation $L(u)$ we see that $L(c) = \text{Id}$ if c is the unit element. If A is commutative c is the unit element if and only if

$$L(c) = \text{Id}.$$

For algebras $A = (X, \circ)$ we define the powers of the elements $u \in A$ by

$$u^1 = u, \ u^{m+1} := u \circ u^m \quad \text{for integer} \ m \geq 1.$$

This definition is equivalent to

$$u^{m+k} = L^m(u)u^k \quad \text{for integers} \ m, k \geq 1.$$

It is not always true that $u^m \circ u^k = u^{m+k}$ holds. If $\mathcal{A}$ has a unit element c, then we put $u^0 = c$ and get $u^m = L^m(u)c$.

There is a relation between trilinear forms and algebras. Let $\sigma(x, y)$ be a non-singular symmetric bilinear form on X, and recall that X is finite dimensional. We consider a trilinear form $\lambda(u, v, w)$ on the space X. If we fix one argument of this trilinear form, then we get a bilinear form in the remaining two arguments. Therefore, for given u the form $\lambda(u, v, w)$ is a bilinear form in v and w and there exists a linear transformation $L(u)$ such that

(3) $$\lambda(u, v, w) = \sigma(L(u)v, w).$$

Since $\lambda(u, v, w)$ is a linear form in u, it follows that (2) holds. Hence we can define a composition $u \circ v = L(u)v$ on X and get an algebra $\mathcal{A} = (X, \circ)$.

If the trilinear form $\lambda(u, v, w)$ is symmetric in u, v, w, then we immediately see that the algebra is commutative and that

$$\sigma(u \circ v, w) = \sigma(u, v \circ w)$$

holds. The last formula yields $\sigma(L(v)u, w) = \sigma(u, L(v)w)$. Thus the transformation $L(v)$ is self-adjoint with respect to the bilinear form σ.

Vice versa, it is easy to see that we get a trilinear form λ on X if we start with an algebra and define $\lambda(u, v, w)$ by (3). If $\mathcal{A}$ is commutative and if $L(u)$ is self-adjoint with respect to σ, then the trilinear form λ is symmetric.

An algebra $\mathcal{A} = (X, \circ)$ over the field K is called a *Jordan algebra* if

(JA.1) $\mathcal{A}$ is a commutative algebra.

(JA.2) $x \circ (x^2 \circ y) = x^2 \circ (x \circ y)$ holds for all $x, y \in \mathcal{A}$.

If the composition of $\mathcal{A}$ is described by the linear transformation $L(x)$, then (JA.2) is equivalent to

(JA.2*) $L(x)L(x^2) = L(x^2)L(x)$.

The axiom (JA.2) may be considered a substitute for the associative law. In view of

$$x^4 = x \circ [x \circ x^2] = x \circ [x^2 \circ x] = x^2 \circ [x \circ x] = x^2 \circ x^2$$

Jordan algebras, for example, satisfy

$$x^4 = x^2 \circ x^2,$$

and therefore

$$x^5 = x \circ x^4 = x \circ (x^2 \circ x^2) = x^2 \circ (x \circ x^2) = x^2 \circ x^3.$$

An algebra $\mathcal{A} = (X, \perp)$ over the field K is called a *Lie algebra* if

(LA.1) $x \perp x = 0$.

(LA.2) $x \perp (y \perp z) + y \perp (z \perp x) + z \perp (x \perp y) = 0$.

An immediate consequence of (LA.1) is the identity

$$x \perp y = -y \perp x.$$

In terms of the linear transformation $L(x)$, defined by $L(x)y = x \perp y$, this is equivalent to

(LA.1*) $L(x)x = 0$.

(LA.2*) $L(x \perp y) = L(x)L(y) - L(y)L(x)$.

Now let K be the field $\mathbf{R}$ of the real numbers and let $\mathcal{A} = (X, \circ)$ be a commutative algebra. Let $u = u(x)$ and $v = v(x)$ be differentiable vector-valued functions on an open subset of X. The definition of the operator Δ_x^w (cf. I, §6) immediately leads to

$$\Delta_x^w(u \circ v) = v \circ \Delta_x^w u + u \circ \Delta_x^w v = L(v)\frac{\partial u}{\partial x}w + L(u)\frac{\partial v}{\partial x}w$$

and consequently

$$\frac{\partial(u \circ v)}{\partial x} = L(v)\frac{\partial u}{\partial x} + L(u)\frac{\partial v}{\partial x}.$$

Suppose that $\mathcal{A}$ has a unit element c. Then the definition of the powers of $x \in \mathcal{A}$ implies

$$\frac{\partial x^{m+1}}{\partial x} = L(x)\frac{\partial x^m}{\partial x} + L(x^m), \quad m \geq 0.$$

Thus we get

$$\frac{\partial x^{m+1}}{\partial x} = \sum_{\mu=0}^{m} L^{m-\mu}(x)L(x^\mu), \quad m \geq 0.$$

Especially we have

$$\frac{\partial x^2}{\partial x} = 2L(x),$$

$$\frac{\partial x^3}{\partial x} = 2L^2(x) + L(x^2),$$

$$\frac{\partial x^4}{\partial x} = 2L^3(x) + L(x)L(x^2) + L(x^3).$$

In the next section we will consider the exponential function of a Jordan algebra $\mathcal{A}$ with unit element c:

$$e(x) = \sum_{m=0}^{\infty} \frac{1}{m!} x^m.$$

The representation

$$e(x) = e^{L(x)} c$$

shows that $e(x)$ is absolutely and locally uniformly convergent for $x \in \mathcal{A}$. Therefore, the mapping $\mathcal{A} \to \mathcal{A}, x \mapsto e(x)$, is real-analytic, because $x \mapsto L(x)$ is real analytic. We have

(4)
$$\frac{\partial e(x)}{\partial x} = \mathrm{Id} + L(x) + \frac{1}{6}(2L^2(x) + L(x^2)) +$$
$$+ \frac{1}{24}(2L^3(x) + L(x)L(x^2) + L(x^3)) + \dots .$$

§5. ω–domains and Jordan algebras

Let (Y, ω, c) be an ω–domain in the vector space X over $\mathbf{R}$ and let $\sigma(x, y)$ be the non-singular bilinear form associated with (Y, ω, c). From the results in §4 we know that we get a commutative algebra on X from any symmetric trilinear form on X. Here we let

$$\lambda(u, v, w) = \frac{1}{2} \Delta_y^u \Delta_y^v \Delta_y^w \log \omega(y)\big|_{y=c}.$$

Since $\omega(y)$ is real-analytic on Y, the form $\lambda(u, v, w)$ is symmetric in all arguments. It follows from the definition of the operator Δ that $\lambda(u, v, w)$ is a trilinear form. We denote the algebra given by $\lambda(u, v, w)$ by $\mathcal{A} = (X, \circ)$. The product in $\mathcal{A}$ is defined by

$$u \circ v = L(u)v,$$

where the linear transformation $L(u)$ is given by

$$\lambda(u, v, w) = \sigma(L(u)v, w).$$

Since λ is symmetric, the product is commutative. We may describe the transformation $L(u)$ and hence the bilinear form in a different way. Recall that

$$-\Delta_y^v \Delta_y^w \log \omega(y) = \sigma(H(y)v, w)$$

holds. This implies that

$$H(y) = \mathrm{Id} - 2L(y - c) + O(|y - c|^2)$$

is the expansion of $H(y)$ in a neighborhood of the point c. To verify this, use

$$\sigma(H(c+\tau x)v,w) = -\Delta_y^w \Delta_y^v \log \omega(y)\big|_{y=c+\tau x}$$

$$= \sigma(v,w) - \tau \Delta_y^x \Delta_y^w \Delta_y^v \log \omega(y)\big|_{y=c} + O(\tau^2).$$

There is an important connection between this algebra $\mathcal{A}$ and the geometric problem of determining the geodesics of the ω–domain. Indeed we will prove that the geodesics through the point c can be given in the form $e(\tau u)$, $\tau \in \mathbf{R}$, where $e(u)$ is the exponential function of $\mathcal{A}$.

We know from Theorem 3 that there is a real-analytic map $X \to Y, x \mapsto g(x)$, such that

$$g(0) = c \quad \text{and} \quad \frac{\partial g(x)}{\partial x}\bigg|_{x=0} = \text{Id}.$$

The curves $g(\tau u)$ are exactly the geodesics through c. Let

$$g(x) = \sum_{m=0}^{\infty} \frac{1}{m!} b_m(x), \quad b_0(x) = c, \quad b_1(x) = x,$$

where the $b_m(x)$ are homogeneous polynomials of degree m. In order to substitute this expansion into the fundamental equation for $g(x)$ (see Theorem 3(c)) we compute

$$g(-\tfrac{\xi}{2}u) = c - \tfrac{\xi}{2}u + \dots, \quad H(g(-\tfrac{\xi}{2}u)) = \text{Id} + \xi L(u) + \dots .$$

Substituting this expansion in

$$g((\xi+\eta)u) = H(g(-\tfrac{\xi}{2}u))g(\eta u)$$

and comparing the coefficients of $\xi \eta^{m-1}$ we obtain

$$b_m(u) = L(u)b_{m-1}(u).$$

In view of $b_0(u) = c$ we get $b_m(u) = L^m(u)c$. From $b_1(u) = u$ it follows that $u = L(u)c = u \circ c$. Therefore c is the unit element of $\mathcal{A}$. Summarizing we obtain

$$g(u) = e(u) = \sum_{m=0}^{\infty} \frac{1}{m!} u^m.$$

The main result of this chapter is

Theorem 4. *Let (Y,ω,c) be an ω–domain and $\sigma(x,y)$ the associated bilinear form. Then there is exactly one Jordan algebra $\mathcal{A} = (X,\circ)$ with unit element c such that*

(a) $g(x) = e(x) \in Y$ for every $x \in X$.

(b) $e^{\sharp}(x) = e(-x)$ for every $x \in X$.

(c) $\sigma(x \circ y, z) = \sigma(x, y \circ z)$ for all $x, y, z \in X$.

Proof. The algebra $\mathcal{A}$ is uniquely determined by (c) due to the calculations above. Considering Theorem 3 we only need to prove that $\mathcal{A}$ is a Jordan algebra. Let

$$E(x) = \frac{\partial e(x)}{\partial x} = \sum_{m=0}^{\infty} \frac{1}{m!} \frac{\partial x^m}{\partial x}.$$

For a differentiable function $y = y(x) : X \to Y$ we have

$$\frac{\partial y^\sharp}{\partial x} = \frac{\partial y^\sharp}{\partial y} \cdot \frac{\partial y}{\partial x} = -H(y)\frac{\partial y}{\partial x}.$$

Applying this formula to $y = e(x)$ it follows from Theorem 3(b) that $E(-x) = H(e(x))E(x)$ and

$$E^*(x)E(-x) = E^*(-x)E(x).$$

Thus, $E^*(x)E(-x)$ does not change if we replace x by $-x$. Hence in the expansion of this function all terms of odd degree vanish. We compute the terms of third degree using formula (4) in §4, hence

$$L(x)L(x^2) = L(x^2)L(x).$$

$\square$

Remark. We get the same results if we replace the group $\Sigma(Y,\omega)$ by the group of linear transformations

$$\{W \in Hom(X,X); \ WY = Y, \ \omega(Wy) = \lambda(W) \cdot \omega(y)\},$$

where $\lambda(W)$ is a certain non-zero constant.

One can prove that the algebra $\mathcal{A} = (X, \circ)$ is associative and semisimple if and only if the ω–domain (Y, ω, c) is given by

$$Y = \{y = \eta_1 u_1 + \ldots + \eta_n u_n; \ \eta_\nu > 0\}, \quad \omega(y) = \gamma \cdot \eta_1^{\alpha_1} \cdot \ldots \cdot \eta_n^{\alpha_n},$$

where $u_1, \ldots, u_n$ is a suitable basis of X and $\alpha_1, \ldots, \alpha_n$ are real numbers.

Notes

The correspondence between domains of positivity and Jordan algebras (in the sense described in this chapter for ω–domains) was described in [18]. Again it turns out that it is not necessary to assume that the domain of positivity is homogeneous; we only need that all $H(y)$ belong to $\Sigma(Y)$. Homogenity is a consequence of this; see Chapter VI. This indicates that the present construction of a Jordan algebra associated with ω–domains is more satisfactory than Vinberg's [26] way of considering homogeneous spaces.

Editors' Notes

1. (i) The following observation concerning the proof of Lemma 1 may be useful: If W is any invertible linear map, then $\omega^*(y) := \omega(Wy)$ is defined for all $y \in W^{-1}Y$, and the chain rule implies $\Delta^u_y \log \omega^*(y) = \Delta^{Wu}_{Wy} \log \omega(Wy)$. For $W \in \Sigma$ it follows that $\Delta^u_y \log \omega(y) = \Delta^{Wu}_{Wy} \log \omega(Wy)$. A similar observation applies to the second derivatives.

(ii) In §2 (B) the identity $\det(e + \tau y + O(\tau^2)) = 1 + \tau \operatorname{tr} y + O(\tau^2)$ is used. It will also be used in the following chapters.

(iii) The geodesics of Y are geodesic with respect to the unique torsion-free affine connection given in coordinates by

$$k(y,z) = H(y)^{-1} \Big(\frac{1}{2} \big(\frac{\partial \sigma(H(y)z,z)}{\partial y} \big)^* - \big(\frac{\partial H(y)}{\partial y} z \big) z \Big).$$

This is intimately connected with the structure of the Jordan algebra associated to the ω–domain. For instance, one has $k(c, z) = z^2$, and generally $k(y, z)$ is given by the square in a certain mutation (see the following chapters).

2. Homogeneous open cones that are not necessarily convex are of interest in applications, as shown by recent work of Faraut and Gindikin [FGi], Gindikin [Gi2], [Gi3], Massam and Neher [MNe1], Neher [Ne2], and Hagenbach and Upmeier [HU], for instance. An algebraic version of this chapter can be found in Braun and Koecher [BK], Kap. 6, §8. The results (with the exception of those on geodesics) can be carried over to this algebraic setting. Later, Dorfmeister and Koecher [DK1] investigated relative invariants of regions in $\mathbf{R}^n$, and their automorphism groups. In this paper, neither homogeneity nor symmetry is assumed from the beginning. Sato and Kimura [SKi] investigated and classified prehomogeneous vector spaces over $\mathbf{C}$, using representation theory. (Complexification of homogeneous open cones naturally leads to such spaces!) A more recent paper on this topics is Müller, Rubenthaler and Schiffmann [MRS]. In [GrW], some results on "ω–domains without (D3)" can be found. Moreover, there are other conditions which ensure that such a domain is related to a Jordan algebra. One of them is that $y \mapsto y^\#$ is involutory for every base point.

Chapter III. Jordan Algebras

§1. Jordan algebras

Let K be a field of a characteristic $\neq 2$. From Chapter II, §4 we recall the definition that a K-vector space A together with a composition $A \times A \to A, (u, v) \mapsto u \circ v$, is called a *Jordan algebra* if all $u, v, w \in A$, $\alpha \in K$ satisfy

(JA.1)   (a) $u \circ v = v \circ u$,

          (b) $u \circ (v + w) = u \circ v + u \circ w$, $(\alpha u) \circ v = \alpha(u \circ v)$,

(JA.2) $u \circ (u^2 \circ v) = u^2 \circ (u \circ v)$, where $u^2 = u \circ u$.

We have seen that there is a linear transformation $L(u)$, which is linear in u, such that $u \circ v = L(u)v$. The axioms (JA.1) and (JA.2) are equivalent to

(JA.1*)   (a) $L(u)v = L(v)u$,

          (b) $L(u)$ is linear in u,

(JA.2*) $L(u)L(u^2) = L(u^2)L(u)$.

In (JA.2) we replace u by $u + \lambda w$, $\lambda \in K$. Comparing the coefficients of λ we get the so-called *polarization formula*:

$$2(u \circ v) \circ (u \circ w) + u^2 \circ (v \circ w) = 2u \circ (v \circ (u \circ w)) + (u^2 \circ v) \circ w.$$

Considering this equation as a linear transformation in w we obtain

(1) $2L(u \circ v)L(u) + L(u^2)L(v) = 2L(u)L(v)L(u) + L(u^2 \circ v)$.

Considering the polarization formula as a linear transformation in v and then substituting v for w we get

(2) $2L(u \circ v)L(u) + L(u^2)L(v) = 2L(u)L(u \circ v) + L(v)L(u^2).$

Two elements $u, v \in \mathcal{A}$ are said to *commute* if

$$u \circ (v \circ x) = v \circ (u \circ x) \quad \text{for every} \quad x \in \mathcal{A}.$$

Thus, u and v commute if and only if the transformations $L(u)$ and $L(v)$ commute. In this notation a commutative algebra is a Jordan algebra if and only if u and u^2 commute for every $u \in \mathcal{A}$. Thus (2) immediately leads to

Lemma 1. *In a Jordan algebra u and $u \circ v$ commute if and only if u^2 and v commute.*

Recall that the powers of $u \in \mathcal{A}$ are defined by

$$u^1 = u, \ u^{m+1} = u \circ u^m \quad \text{for} \quad m \geq 1.$$

In order to prove the law of associativity for the powers of a given element $u \in \mathcal{A}$, we first show

Theorem 1. *If $\mathcal{A}$ is a Jordan algebra then $L(u^m)$ is a polynomial in $L(u)$ and $L(u^2)$.*

Proof. The claim is trivial for $m = 1$ and 2. Assume that $L(u^r)$ is a polynomial in $L(u)$ and $L(u^2)$ for $1 \leq r \leq m$. Then u^r and u commute and we get

$$u^{m-1} \circ (u \circ u) = u \circ (u^{m-1} \circ u) = u \circ u^m = u^{m+1}$$

or $u^2 \circ u^{m-1} = u^{m+1}$. Substituting u^{m-1} for v in (1) yields

(∗) $2L(u^m)L(u) + L(u^2)L(u^{m-1}) = 2L(u)L(u^{m-1})L(u) + L(u^{m+1}).$

Hence, by induction $L(u^{m+1})$ is a polynomial in $L(u)$ and $L(u^2)$, and the claim is proved.   □

Corollary 1. *In a Jordan algebra the powers u^r and u^s commute for $r \geq 1$ and $s \geq 1$.*

Proof. If we apply Theorem 1 to $L(u^r)$ and $L(u^s)$, we see that both transformations are polynomials in $L(u)$ and $L(u^2)$. Since $L(u)$ and $L(u^2)$ commute, this is also true for $L(u^r)$ and $L(u^s)$.   □

We note that equation (∗) yields

Corollary 2. *In a Jordan algebra one has*

$$L(u^{m+1}) = 2L(u^m)L(u) - L(u^{m-1})(2L^2(u) - L(u^2)), \quad m \geq 2.$$

Now we show

(3) $u^r \circ (u^s \circ v) = u^{r+s} \circ v$ for $r, s \geq 1$ provided that u and v commute.

This is trivial if $r + s = 2$. For $s = 1$ we apply Corollary 1 and get

$$u^r \circ (u \circ v) = u \circ (u^r \circ v) = v \circ (u \circ u^r) = v \circ u^{r+1}.$$

Let us assume that (3) is true for $r + s \leq m$. If $r + s = m + 1$ and $s > 1$ we obtain

$$u^r \circ (u^s \circ v) = u^r \circ (v \circ (u^{s-1} \circ u)) = u^r \circ (u \circ (u^{s-1} \circ v))$$
$$= u \circ (u^r \circ (u^{s-1} \circ v)) = u \circ (u^{r+s-1} \circ v)$$
$$= v \circ (u^{r+s-1} \circ u) = v \circ u^{r+s}.$$

Hence (3) holds.

In (3) we substitute $v = u$ and get

Theorem 2. *If A is a Jordan algebra, then $u^r \circ u^s = u^{r+s}$ holds for all $u \in A$, $r \geq 1$ and $s \geq 1$.*

This theorem shows that the associative law holds for the powers of each element in A. Thus A is a *power-associative* algebra.

Given a polynomial $f(\tau) = \alpha_1\tau + \ldots + \alpha_m\tau^m \in K[\tau]$ in τ with constant term zero, we put $f(u) = \alpha_1 u + \ldots + \alpha_m u^m$ for $u \in A$. Given $u \in A$ let $K_1[u]$ be the vector space in A spanned by $u^1, u^2, \ldots$. The elements of $K_1[u]$ are those of the form $f(u)$, where $f(\tau)$ is an arbitrary polynomial over K with zero constant term. $K_1[u]$ is a subspace of A and Theorem 2 yields that $K_1[u]$ is an associative subalgebra of A if A is a Jordan algebra. It follows from (3) that

(4) $x \circ (y \circ v) = (x \circ y) \circ v$ for $x, y \in K_1[u]$ provided that u and v commute.

We consider the vector space $K_1[x, y]$ spanned over K by all the terms x^r, y^s and $x^r \circ y^s$ for $r, s \geq 1$. Under which conditions is $K_1[x, y]$ an associative subalgebra of A?

Theorem 3. *Let A be a Jordan algebra. Given $x, y \in A$ the following assertions are equivalent.*

(i) $x, y, x \circ y$ commute pairwise.

(ii) $K_1[x, y]$ is an associative subalgebra of A whose elements commute pairwise.

Proof. Since $x, y, x \circ y$ belong to $K_1[x, y]$, the implication "(ii) $\Rightarrow$ (i)" is trivial. The proof of the implication "(i) $\Rightarrow$ (ii)" is done in 3 steps:

(a) *All the terms* $x^r, y^s, x \circ y$ *commute by pairs and we have*

$$y^q \circ (x^r \circ y^s) = x^r \circ y^{q+s}, \quad x^p \circ (x^r \circ y^s) = x^{p+r} \circ y^s.$$

Applying Lemma 1 to $u = x$, $v = y$, we conclude that x^2, y commute and analogously y^2, x. Substituting $u = x$, $v = y$ in (1) we see that $x^2 \circ y, x, y$ commute by pairs. Applying Lemma 1 to $u = y$, $v = x^2$ we get that x^2, y^2 commute. Applying Lemma 1 to $u = x$, $v = x \circ y$ we get that $x^2, x \circ y$ commute; analogously $y^2, x \circ y$ commute. Hence the claim is true for $r, s = 1, 2$. Theorem 1 shows that $x^r, y^s, x \circ y$ commute by pairs for every r and s. Therefore

$$y^q \circ (x^r \circ y^s) = x^r \circ (y^q \circ y^s) = x^r \circ y^{q+s}.$$

(b) $L(x^r \circ y^s)$ *is a polynomial in* $L(x), L(x^2), L(y), L(y^2), L(x \circ y)$.

We prove the assertion for $r = 1$ by induction on s, putting $x \circ y^s = x$ for $s = 0$. Assume that (b) is true for $L(x \circ y^{s-1})$ and $L(x \circ y^s)$. Substituting $u = y$, $v = x \circ y^{s-1}$ in (1) we get the result for $s + 1$.

Now we prove (b) for a given s by induction on r. Assume that $L(x^{r-1} \circ y^s)$ and $L(x^r \circ y^s)$ are polynomials in $L(x), \dots, L(x \circ y)$. Substituting $u = x$, $v = y^s \circ x^{r-1}$ in (1), we get the assertion for $r + 1$.

(c) *Proof of the theorem.*

The hypothesis and (b) imply that $x^r, x^p \circ y^q$ commute. Thus (a) yields

$$(x^r \circ y^s) \circ (x^p \circ y^q) = x^r \circ ((x^p \circ y^q) \circ y^s) = x^r \circ (x^p \circ y^{q+s}) = x^{r+p} \circ y^{q+s}.$$

$K_1[x, y]$ is therefore closed with respect to multiplication and is associative.

$\square$

An important example of an associative subalgebra of $\mathcal{A}$ is the *center* $Z(\mathcal{A})$ of $\mathcal{A}$, which consists of all elements $z \in \mathcal{A}$ such that z, u commute for any $u \in \mathcal{A}$. As an equation this means, for $z \in Z(\mathcal{A})$ and $u, v \in \mathcal{A}$,

$$u \circ (z \circ v) = z \circ (u \circ v) = v \circ (z \circ u).$$

This easily implies that $Z(\mathcal{A})$ is an associative algebra. By definition z belongs to $Z(\mathcal{A})$ if and only if $L(z)$ commutes with all $L(u)$, $u \in \mathcal{A}$. Note that the equation $u \circ (z \circ v) = (z \circ u) \circ v$ implies

$$L(z \circ u) = L(z)L(u), \quad z \in Z(\mathcal{A}), \ u \in \mathcal{A}.$$

Lemma 2. *If A is a Jordan algebra with unit element c and if $A : A \to A$ is a linear transformation, then $AL(u) = L(u)A$ for every $u \in A$ implies $A = L(z)$ for some $z \in Z(A)$.*

Proof. By hypothesis we obtain $A(u \circ v) = u \circ Av$ for every $u, v \in A$. This implies $AL(v) = L(Av)$. Substituting $v = c$, $z = Ac$, we get $A = L(z)$. The hypothesis now shows that $z \in Z(A)$. □

Later on we need an important transformation $P(u)$, which is defined for every commutative algebra A as follows:

$$P(u)v = 2u \circ (u \circ v) - u^2 \circ v.$$

As $u \circ v = L(u)v$ we obtain

$$P(u) = 2L^2(u) - L(u^2).$$

If A contains a unit element c, then $L(c) = \text{Id}$ and $P(c) = \text{Id}$ follow. The linear transformation $P(u)$ is called the *quadratic representation of A*.

Theorem 4. *Let A be a commutative algebra over K with unit element c, and $\text{char } K \neq 2, 3, 5$. Then A is a Jordan algebra if and only if $P(u^2) = P^2(u)$ holds for every $u \in A$.*

Proof. "$\Rightarrow$": Corollary 2 immediately yields

$$L(u^4) = 4L(u^2)L^2(u) - 4L^4(u) + L^2(u^2)$$

and this gives $P(u^2) = P^2(u)$.

"$\Leftarrow$": We substitute $u = c + \lambda x$, $\lambda \in K$, in $P(u^2) = P^2(u)$ and compare the coefficients of λ^3. This implies

$$4L^3(x) + 2L(x^3) = 3L(x)L(x^2) + 3L(x^2)L(x).$$

Applying the assumption to c we get $u^4 = u^2 \circ u^2$. Here we substitute $u = x + \lambda y$ and compare the coefficients of λ. We obtain

$$4x^2 \circ (x \circ y) = 2x \circ (x \circ (x \circ y)) + x \circ (x^2 \circ y) + x^3 \circ y,$$
$$4L(x^2)L(x) = 2L^3(x) + L(x)L(x^2) + L(x^3).$$

The combination of both equations yields $5L(x^2)L(x) = 5L(x)L(x^2)$, which is (JA.2*). □

§2. The radical of a Jordan algebra.

We assume for the rest of this chapter that $\mathcal{A}$ is a Jordan algebra over K and that the characteristic of K is zero or otherwise char $K > 2 \dim \mathcal{A}$.

An element $e \in \mathcal{A}$ is called an *idempotent* if $e^2 = e$ and $e \neq 0$. Substituting $u = e = e^2$ and $m = 2$ in Corollary 2 we get the formula

$$2L^3(e) - 3L^2(e) + L(e) = 0,$$

for an idempotent e. The eigenvalues of the transformation $L(e)$ (in the algebraic closure of K) are therefore solutions of $2\tau^3 - 3\tau^2 + \tau = 0$, i.e. are among $0, \frac{1}{2}$ and 1. It also follows that $L(e)$ is diagonalizable. In view of $e = e^2 = L(e)e$, one eigenvalue of $L(e)$ is equal to 1. We have $\mathrm{tr} L(e) \neq 0$, since $\mathrm{tr} L(e)$ is the sum of the eigenvalues (including multiplicities).

An element $v \in \mathcal{A}$ is called *nilpotent* if $v^r = 0$ for some $r \geq 1$. We consider the associative algebra $K_m[u]$ spanned by $u^m, u^{m+1}, \dots$. If u is nilpotent then each element of $K_m[u]$ is nilpotent, too. Vice versa we have

Lemma 3. *If u is not nilpotent then there exist infinitely many integers m such that $K_m[u]$ contains an idempotent.*

Proof. It is sufficient to show that there is an idempotent e in some $K_m[u]$. Then we have $e = e^2 \in K_{2m}[u]$. In view of

$$\mathcal{A} \supset K_1[u] \supset K_2[u] \supset \dots \supset K_m[u] \supset \dots$$

and $K_m[u] \neq \{0\}$ there exists a number m such that

$$u^m \circ K_m[u] = K_{2m}[u] = K_m[u].$$

Since $K_m[u]$ has finite dimension, the map

$$K_m[u] \to K_m[u], \; v \mapsto u^m \circ v,$$

is a bijective endomorphism of the vector space $K_m[u]$. Hence there exists $0 \neq e \in K_m[u]$ such that $u^m = u^m \circ e$, and therefore $u^l = u^l \circ e$ for all $l \geq m$, and $e^2 = e$.   □

Theorem 5. *Given $u \in \mathcal{A}$ the following three statements are equivalent:*

(A) u is nilpotent.

(B) $L(u)$ is nilpotent.

(C) There is a positive integer m, such that $\mathrm{tr} L(u^r) = 0$ for every $r \geq m$.

If one of these statements is true then (C) holds for $m = 1$.

Proof. "(C) $\Rightarrow$ (A)": Assume that u is not nilpotent. We find $m \geq 1$ such that (C) holds and by Lemma 3 there is an idempotent $e \in K_m[u]$. Then $\mathrm{tr}L(v)$ vanishes for every $v \in K_m[u]$ and therefore $\mathrm{tr}L(e) = 0$. This is a contradiction.

"(A) $\Rightarrow$ (B)": Corollary 2 yields the identity

(1) $$L(u^3) = 3L(u^2)L(u) - 2L^3(u).$$

Now we show for $r \geq 2$ that $u^r = 0$ implies nilpotency of $L(u)$. By (1), this is true for $r = 2$. Let $u^{r+1} = 0$ and assume that the statement holds for r. From $(u^2)^r = (u^3)^r = 0$ we conclude that $L(u^2)$ and $L(u^3)$ are nilpotent. Then (1) shows that $L^3(u)$ is a sum of two nilpotent transformations which commute. Hence $L(u)$ is nilpotent.

"(B) $\Rightarrow$ (C)": For any $r \geq 1$ the element u^r is nilpotent and therefore $L(u^r)$ is nilpotent. Since the trace of a nilpotent linear transformation is zero, the theorem is proved. $\qquad\square$

Let $\mathcal{A}$ and $\mathcal{A}'$ be two algebras over K. A homomorphism of the vector spaces $\phi : \mathcal{A} \to \mathcal{A}'$ is called a *homomorphism of algebras* if $\phi(u \circ v) = \phi(u) \circ \phi(v)$ for all $u, v \in \mathcal{A}$.

A subset $\mathcal{I}$ of $\mathcal{A}$ is called an *ideal* of an algebra $\mathcal{A}$ if

(I.1) $\mathcal{I}$ is a subspace of the vector space $\mathcal{A}$.

(I.2) $u \circ d \in \mathcal{I}$ and $d \circ u \in \mathcal{I}$ for all $u \in \mathcal{A}$, $d \in \mathcal{I}$.

The kernel of a homomorphism is therefore an ideal. The ideal $\mathcal{I}$ defines a congruence relation on $\mathcal{A}$ by

$$u \equiv v \ (\mathrm{mod}\ \mathcal{I}) \ \Leftrightarrow \ u - v \in \mathcal{I}.$$

Let $\mathcal{A}/\mathcal{I}$ be the set of all congruence classes $\bar{u} = u \ (\mathrm{mod}\ \mathcal{I})$ for $u \in \mathcal{A}$. In the usual way $\mathcal{A}/\mathcal{I}$ becomes an algebra over K. If $\mathcal{A}$ is a Jordan algebra then $\mathcal{A}/\mathcal{I}$ is a Jordan algebra. Since $\mathcal{I}$ is the kernel of the map $u \to \bar{u}$, a set $\mathcal{I} \subset \mathcal{A}$ is an ideal of $\mathcal{A}$ if and only if $\mathcal{I}$ is the kernel of a homomorphism $\phi : \mathcal{A} \to \mathcal{A}'$ for some algebra $\mathcal{A}'$.

In the following let $\mathcal{A}$ again be a Jordan algebra. Given $u, v \in \mathcal{A}$ we define

$$\tau(u, v) := \mathrm{tr}L(u \circ v).$$

Obviously $\tau(u, v)$ is a symmetric bilinear form on the vector space $\mathcal{A}$.

Lemma 4. $\tau(u \circ v, w) = \tau(u, v \circ w)$ *holds for* $u, v, w \in \mathcal{A}$.

Proof. We start with the polarization formula and substitute $u = x + y$, $v = z$. Comparing the terms which are linear in x and y, we get

(2)
$$(x \circ z) \circ (y \circ w) + (y \circ z) \circ (x \circ w) + (x \circ y) \circ (z \circ w)$$
$$= x \circ (z \circ (y \circ w)) + y \circ (z \circ (x \circ w)) + w \circ (z \circ (x \circ y)).$$

This is a linear transformation in w. Thus one sees

(3)
$$L(z \circ x)L(y) + L(y \circ z)L(x) + L(x \circ y)L(z)$$
$$= L(x)L(z)L(y) + L(y)L(z)L(x) + L(z \circ (x \circ y)).$$

A cyclic permutation yields

$$L(y \circ z)L(x) + L(x \circ y)L(z) + L(z \circ x)L(y)$$
$$= L(z)L(y)L(x) + L(x)L(y)L(z) + L(y \circ (z \circ x)).$$

Hence, we obtain

$$\tau(y \circ x, z) - \tau(y, x \circ z) = \mathrm{tr}\big[L((y \circ x) \circ z) - L(y \circ (x \circ z))\big]$$
$$= \mathrm{tr}\big[L(z)L(y)L(x) + L(x)L(y)L(z) - L(x)L(z)L(y) - L(y)L(z)L(x)\big] = 0,$$

in view of $\mathrm{tr}(AB) = \mathrm{tr}(BA)$.   □

We define the *radical* of the Jordan algebra A by

$$rad\, A = \{a \in A;\ \tau(a, u) = 0 \quad \text{for every } u \in A\}.$$

We call the Jordan algebra A *semisimple* if $rad\, A = \{0\}$, i.e. if $\tau(u, v)$ is non-singular. Lemma 4 implies that $rad\, A$ is an ideal in A. The radical may be described without using the bilinear form.

Theorem 6. *Given $a \in A$ one has $a \in rad\, A$ if and only if $a \circ x$ is nilpotent for every $x \in A$.*

Proof. "$\Rightarrow$": From $\tau(a, a^r) = 0$ for all r we get $\mathrm{tr}L(a^m) = 0$ for $m \geq 2$. Theorem 5 shows that a is nilpotent. Since $rad\, A$ is an ideal, we may replace a by $a \circ x$, $x \in A$.

"$\Rightarrow$": Let $a \circ x$ be nilpotent. Then Theorem 5 implies $\mathrm{tr}L(a \circ x) = 0$ and $\tau(a, x) = 0$ for every $x \in A$, hence $a \in rad\, A$.   □

Using Theorem 6 and power-associativity, one may prove that, for any Jordan algebra A, the factor algebra $A/rad\, A$ is semisimple.

Theorem 7. *Let I be an ideal in the Jordan algebra A, then*

$$rad\, I = I \cap rad\, A.$$

Corollary 3. *Every ideal in a semisimple Jordan algebra A is a semisimple Jordan algebra.*

Proof. (a) The inclusion $\mathcal{I} \cap rad\, A \subset rad\, \mathcal{I}$ is obvious.

(b) We prove that $u \in rad\, \mathcal{I}$, $u^2 \in rad\, A$ imply $u \in rad\, A$. Substituting $v = w$ in the polarization formula we obtain

$$2(u \circ v)^2 = 2u \circ (v \circ (u \circ v)) + (u^2 \circ v) \circ v - u^2 \circ v^2.$$

Now one has

$$u \circ (v \circ (u \circ v)) \in u \circ (v \circ \mathcal{I}) \subset u \circ \mathcal{I} \subset rad\, \mathcal{I},$$
$$(u^2 \circ v) \circ v \in (rad\, A \circ v) \circ v \cap (\mathcal{I} \circ v) \circ v \subset rad\, A \cap \mathcal{I} \subset rad\, \mathcal{I},$$
$$u^2 \circ v^2 \in rad\, A \circ v^2 \cap \mathcal{I} \circ v^2 \subset rad\, A \cap \mathcal{I} \subset rad\, \mathcal{I},$$

and therefore $(u \circ v)^2 \in rad\, \mathcal{I}$. Applying Theorem 6 to $\mathcal{I}$ we see that $u \circ v$ is nilpotent. This holds for every $v \in A$ and Theorem 6 implies $u \in rad\, A$.

(c) Given $u \in rad\, \mathcal{I}$ there is an integer m such that $u^{2^m} = 0 \in rad\, A$. The application of (b) yields $u^{2^{m-1}} \in rad\, A$. Repeating this process we finally get $u \in rad\, A$.

□

§3. The unit element of a Jordan algebra

In general a Jordan algebra does not have a unit element, but there exists an extension $\tilde{A}$ of A which possesses a unit element. Indeed, let Kc be a one-dimensional vector space over K with basis element c and let $\tilde{A}$ be the direct sum of A and Kc consisting of the elements $u + \alpha c$, with $u \in A$ and $\alpha \in K$. We define a product on $\tilde{A}$ by

$$(u + \alpha c) \circ (v + \beta c) := (u \circ v + \alpha v + \beta u) + \alpha \beta c.$$

It is easy to see that $\tilde{A}$ is a Jordan algebra, which contains (an isomorphic copy of) A. Obviously c is the unit element of $\tilde{A}$.

In this section we will prove that any semisimple Jordan algebra has a unit element. For this one has to consider a certain decomposition of the vector space A. Let us assume that there is an idempotent e in A. Putting

$$\psi(\tau) = \tau(\tau - 1)(2\tau - 1) = 2\tau^3 - 3\tau^2 + \tau,$$

we know from §2 that $\psi(L(e)) = 0$. One introduces the subspaces

$$A_\lambda := A_\lambda(e) := \{x \in A; \ e \circ x = \lambda x\} \quad \text{for} \quad \lambda = 0, 1/2, 1.$$

The sum of the polynomials

$$\psi_0(\tau) = (\tau - 1)(2\tau - 1) = 2\tau^2 - 3\tau + 1,$$
$$\psi_{1/2}(\tau) = -4\tau(\tau - 1) = -4\tau^2 + 4\tau,$$
$$\psi_1(\tau) = \tau(2\tau - 1) = 2\tau^2 - \tau$$

is 1, and $(\tau - \lambda)\psi_\lambda(\tau)$ is a multiple of $\psi(\tau)$ for all λ. Let

$$x_\lambda = \psi_\lambda(L(e))x.$$

This yields a unique decomposition $x = x_0 + x_{1/2} + x_1$, $x_\lambda \in A_\lambda$, for any $x \in A$. Therefore, A is the direct sum of the subspaces

$$A = A_0 + A_{1/2} + A_1.$$

Theorem 8. *Let e be an idempotent of A and let $A_\lambda = A_\lambda(e)$ for $\lambda = 0, 1/2, 1$. Then:*

(a) A_0 and A_1 are subalgebras of A satisfying

$$A_0 \circ A_1 = \{0\} \quad \text{and} \quad rad\, A_\lambda = A_\lambda \cap rad\, A \quad \text{for} \quad \lambda = 0, 1.$$

(b) $A_0 \circ A_{1/2} \subset A_{1/2}$, $A_1 \circ A_{1/2} \subset A_{1/2}$, $A_{1/2} \circ A_{1/2} \subset A_0 + A_1$.

(c) $\tau(x, y) = 0$ for $x \in A_\lambda$, $y \in A_\mu$, whenever $\lambda \neq \mu$.

Proof. Substituting $u = e$, $e \circ v = \lambda v$, $e \circ w = \mu w$ in the polarization formula, leads to

$$(1 - 2\mu)[e \circ (v \circ w) - \lambda(v \circ w)] = 0, \quad v \in A_\lambda, \ w \in A_\mu.$$

For $\lambda = \mu = 0$ or $\lambda = \mu = 1$ we see that A_0 and A_1 are subalgebras of A. For $\lambda = 0$, $\mu = 1$ resp. $\lambda = 1$, $\mu = 0$ it follows that $A_0 \circ A_1 = \{0\}$. In the case $\mu = 0$ or 1, $\lambda = 1/2$, we get the first two inclusions of (b).

Substituting $v = w = e$, $u \in A_\lambda$ in the polarization formula, we conclude

$$e \circ (e \circ u^2) = e \circ u^2, \quad u \in A_\lambda.$$

We apply this formula to $u = x + \xi y$, $x, y \in A_{1/2}$, and compare the linear terms in ξ: $e \circ (e \circ (x \circ y)) = e \circ (x \circ y)$. Let $x \circ y = a_0 + a_{1/2} + a_1$ be the direct sum decomposition of $x \circ y$, then we get $\frac{1}{4}a_{1/2} + a_1 = \frac{1}{2}a_{1/2} + a_1$. Hence $a_{1/2} = 0$ and $x \circ y \in A_0 + A_1$ follow.

For $\lambda \neq \mu$ and $u \subset A_\lambda$, $v \in A_\mu$ Lemma 4 yields

$$\lambda \cdot \tau(u, v) = \tau(e \circ u, v) = \tau(u, e \circ v) = \mu \cdot \tau(u, v),$$

and this completes the proof of (c).

Theorem 6 implies that $A_\lambda \cap rad\ A \subset rad\ A_\lambda$ for $\lambda = 0, 1$. Let $u = u_\lambda \in rad\ A_\lambda$ and $v = v_0 + v_{1/2} + v_1 \in A$. Then we get $\tau(u, v) = \tau(u, v_\lambda) = 0$, hence $u \in rad\ A$. $\qquad\square$

We will use this decomposition of A with respect to an idempotent in the proof of the following

Theorem 9. *Every semisimple Jordan algebra has a unit element.*

The *proof* is divided into several parts. An idempotent e is called *maximal* if there is no idempotent f in A satisfying $e \circ f = 0$. We may assume $A \neq \{0\}$.

(a) *There exists an idempotent in A, whenever $rad\ A \neq A$.*

In view of $rad\ A \neq A$ it follows from Theorem 6 that there is a non-nilpotent $u \in A$. Lemma 3 shows that there is an idempotent in $K_m[u] \subset A$ for some m.

(b) *If $rad\ A \neq A$ then there is a maximal idempotent e in A.*

In the set of all idempotents of A we choose an idempotent e such that the dimension of the vector space $A_0 = A_0(e)$ is minimal. If there is an idempotent f with $e \circ f = 0$, then $e + f$ is also an idempotent. We show that

$$(1) \qquad A_0(e + f) \subset A_0(e), \quad A_0(e + f) \neq A_0(e).$$

Indeed, for $u \in A_0(e + f)$ we get $u \circ e + u \circ f = 0$. By $f \in A_1(e + f)$ we conclude from Theorem 8(a) that $u \circ f = 0$ and therefore $u \circ e = 0$, hence $u \in A_0(e)$. In view of $f \circ e = 0$ and $f \circ (e + f) = f \neq 0$ we get $f \in A_0(e)$, but $f \notin A_0(e + f)$. Since (1) contradicts the choice of e, there exists no such idempotent, and therefore e is maximal.

(c) $A_0(e) = \{0\}$.

$A_0(e)$ does not contain an idempotent, as was seen in (b). Part (a) yields that every $u \in A_0$ is nilpotent, hence $A_0(e) = rad\ A_0(e)$; and Theorem 8(a) gives $A_0(e) \subset rad\ A$. Since $A(e)$ is semisimple we get $A_0(e) = \{0\}$.

(d) $A_{1/2}(e) = \{0\}$.

Let $u = u_{1/2} \in A_{1/2}(e)$ and $v = v_{1/2} + v_1 \in A$. Theorem 8(c) implies

$$\tau(u, v) = \tau(u, v_{1/2}) = \mathrm{tr}L(u \circ v_{1/2}).$$

It follows from (c) and Theorem 8(b) that $a := u \circ v_{1/2} \in A_1(e)$. Therefore we get

$$\tau(u,v) = \mathrm{tr}L(a) = \mathrm{tr}L(e \circ a) = \tau(e,a)$$
$$= \tau(e \circ u, v_{1/2}) = \frac{1}{2}\tau(u, v_{1/2}) = \frac{1}{2}\tau(u,v).$$

Thus $\tau(u,v) = 0$ holds for every $v \in A$, hence $u = 0$, because of $\mathrm{rad}A = \{0\}$.

(e) It follows from (c) and (d) that $A = A_1(e)$. Therefore e is the unit element of A.   □

We know from Lemma 4 that the bilinear form $\tau(u,v) = \mathrm{tr}L(u \circ v)$ satisfies $\tau(u \circ v, w) = \tau(u, v \circ w)$. Given a semisimple Jordan algebra $\tau(u,v)$ is non-singular by definition. In this case we can describe all bilinear forms satisfying the latter condition:

Theorem 10. *Let A be a Jordan algebra with unit element c. If there exists a non-singular symmetric bilinear form $\mu(u,v)$ on A satisfying $\mu(u \circ v, w) = \mu(u, v \circ w)$, then the following statements are equivalent:*

(i) $\sigma(u,v)$ is a symmetric bilinear form on A satisfying $\sigma(u \circ v, w) = \sigma(u, v \circ w)$ for every $u, v, w \in A$.

(ii) There exists an element z in the center $Z(A)$ of A such that $\sigma(u,v) = \mu(z \circ u, v)$.

Proof. "(ii) $\Rightarrow$ (i)": The claim follows immediately from the definition of the center.

"(i) $\Rightarrow$ (ii)": Since $\mu(u,v)$ is non-singular, there exists a linear transformation T such that

$$\sigma(u,v) = \mu(Tu, v).$$

We get

$$\mu(T(v \circ u), w) = \mu(T(u \circ v), w) = \sigma(u \circ v, w) = \sigma(u, v \circ w)$$
$$= \mu(Tu, v \circ w) = \mu(Tu \circ v, w) = \mu(v \circ Tu, w)$$

and therefore $T(v \circ u) = L(v)Tu$ or $TL(v) = L(v)T$. Now Lemma 2 shows $T = L(z)$ for some $z \in Z(A)$.   □

§4. The decomposition theorem

A Jordan algebra A is said to be the *direct sum* of the subalgebras $I_1, \ldots, I_r$, symbolically $A = I_1 \oplus \ldots \oplus I_r$, if A is the direct sum of the vector spaces $I_1, \ldots, I_r$, and if $I_\nu \circ I_\mu = \{0\}$ holds for $\nu \neq \mu$. Each subalgebra I_ν is called a *direct summand* of A. It is obvious that any direct summand of A is an ideal in $\mathcal{A}$. The algebra $\mathcal{A}$ is called *indecomposable* if there is no direct sum decomposition of $\mathcal{A}$ with two non-trivial subalgebras. $\mathcal{A}$ is called a *simple* algebra if $\mathcal{A}$ is semisimple and indecomposable. Finally, an ideal $\mathcal{I} \neq \{0\}$ in $\mathcal{A}$ is called *minimal* if $\mathcal{J} \subset \mathcal{I}$ for an ideal $\mathcal{J}$ of $\mathcal{A}$ implies $\mathcal{J} = \{0\}$ or $\mathcal{J} = \mathcal{I}$.

If $\mathcal{A}$ is semisimple, we next prove the following statements:

(a) *Any ideal $\mathcal{I} \neq \{0\}$ of $\mathcal{A}$ is a direct summand of $\mathcal{A}$.*

It follows from Corollary 3 that $\mathcal{I}$ is a semisimple Jordan algebra. Thus there exists a unit element e of $\mathcal{I}$. Clearly e is an idempotent in $\mathcal{A}$. We form the decomposition of $\mathcal{A}$ according to §3:

$$A = A_0 + A_{1/2} + A_1.$$

Given $x \in A_\lambda$, $\lambda = \frac{1}{2}$ or 1, we get $x = \frac{1}{\lambda} e \circ x \in \mathcal{I}$. Since e is the unit element of $\mathcal{I}$ we obtain $\lambda = 1$ for $x \neq 0$. Hence, we have $A_{1/2} = \{0\}$ and $A_1 = \mathcal{I}$. It follows that $\mathcal{A} = \mathcal{I} \oplus A_0$, because Theorem 8(a) implies $\mathcal{I} \circ A_0 = \{0\}$.

(b) *If $\mathcal{I} \neq \{0\}$ is an ideal in $\mathcal{A}$ and $\mathcal{J}$ is an ideal in $\mathcal{I}$, then $\mathcal{J}$ is also an ideal in $\mathcal{A}$.*

Using (a) it follows that $\mathcal{I}$ is a direct summand of $\mathcal{A} : \mathcal{A} = \mathcal{I} \oplus \mathcal{I}'$. One has $\mathcal{J} \circ \mathcal{I} \subset \mathcal{J}$ and $\mathcal{J} \circ \mathcal{I}' \subset \mathcal{I} \circ \mathcal{I}' = \{0\}$, hence $\mathcal{J} \circ \mathcal{A} \subset \mathcal{J}$.

(c) *If $\mathcal{J}, \mathcal{I}$ are ideals in $\mathcal{A}$, and $\mathcal{J}$ is minimal, then $\mathcal{J} \subset \mathcal{I}$ or $\mathcal{I} \cap \mathcal{J} = \mathcal{I} \circ \mathcal{J} = \{0\}$.*

$\mathcal{J} \cap \mathcal{I}$ is an ideal of $\mathcal{I}$. Since $\mathcal{J}$ is minimal, we have $\mathcal{J} \cap \mathcal{I} = \mathcal{J}$ or $\mathcal{J} \cap \mathcal{I} = \{0\}$. In the latter case we get $\mathcal{J} \circ \mathcal{I} \subset \mathcal{J} \cap \mathcal{I} = \{0\}$.

(d) *There exists at least one decomposition $\mathcal{A} = \mathcal{I}_1 \oplus \ldots \oplus \mathcal{I}_r$ into simple algebras $\mathcal{I}_\nu$, $\nu = 1, \ldots, r$.*

Indeed, assume that $\mathcal{A}$ is not simple. Then there exists a non-trivial ideal $\mathcal{I}$ in $\mathcal{A}$ and we conclude from (a) that $\mathcal{A} = \mathcal{I} \oplus \mathcal{I}'$. The assertion follows by induction on the dimension.

(e) *In (d) the $\mathcal{I}_\nu$ are exactly the minimal ideals of $\mathcal{A}$.*

Let $\mathcal{I}$ be a minimal ideal of $\mathcal{A}$. It follows from (c) that $\mathcal{I} \subset \mathcal{I}_\nu$ for some ν. (Otherwise $\mathcal{I} \circ \mathcal{I}_\nu = \{0\}$ for all ν, hence $\mathcal{I} \circ \mathcal{A} = \{0\}$ and $\mathcal{I} = \{0\}$.) Conversely, let us assume that one of the $\mathcal{I}_\nu$ in (d) is not a minimal ideal. Then there exists a non-trivial ideal $\mathcal{J}$ of this $\mathcal{I}_\nu$. It follows from (a) that $\mathcal{J}$ is a direct summand of $\mathcal{I}_\nu$, hence $\mathcal{I}_\nu$ is not simple.

Summarizing we have:

Theorem 11. *Every semisimple Jordan algebra $\mathcal{A}$ is a direct sum of simple Jordan algebras. These direct summands are exactly the minimal ideals of $\mathcal{A}$. Two decompositions of a semisimple Jordan algebra into direct sums of simple Jordan algebras are equal up to a permutation of the summands.*

We now prove:

A subset $\mathcal{I} \neq \{0\}$ of $\mathcal{A}$ is an ideal in $\mathcal{A}$ if and only if there is an idempotent e in $Z(\mathcal{A})$ such that $\mathcal{I} = e \circ \mathcal{A}$.

Indeed, if $\mathcal{I} = e \circ \mathcal{A}$, $e \in Z(\mathcal{A})$, then $\mathcal{I}$ is obviously an ideal. Vice versa, let $\mathcal{I}$ be an ideal in $\mathcal{A}$ and let e be the unit element of the algebra $\mathcal{I}$. We obtain $\mathcal{A} = \mathcal{I} \oplus \mathcal{A}_0(e)$, $\mathcal{I} = \mathcal{A}_1(e) = e \circ \mathcal{A}$, $\mathcal{A}_{1/2}(e) = \{0\}$. Given $u = u_0 + u_1$, $v = v_0 + v_1 \in \mathcal{A}$ we get

$$e \circ (u \circ v) - u \circ (e \circ v) = e \circ (u_0 \circ v_0 + u_1 \circ v_1) - u_1 \circ v_1 = 0.$$

Hence $e \in Z(\mathcal{A})$ follows.

Let $\mathcal{A} = \mathcal{I}_1 \oplus \ldots \oplus \mathcal{I}_r$ be a direct sum decomposition and let $\mathcal{I}_\nu = e_\nu \circ \mathcal{A}$, where $e_\nu^2 = e_\nu \in Z(\mathcal{A})$. The unit element c of A is given by $c = e_1 + \ldots + e_r$, and $e_\nu \circ e_\mu = \delta_{\nu\mu} e_\nu$.

§5. The inverse

In this section we assume that the Jordan algebra $\mathcal{A}$ contains a unit element c. Recalling the definition $u^0 := c$, $u \in \mathcal{A}$, we see that $u^r \circ u^s = u^{r+s}$ holds for all integers $r, s \geq 0$. Let $K[u]$ denote the vector space spanned by $c, u, u^2, \ldots$. Then $K[u]$ is an associative subalgebra of $\mathcal{A}$ which consists of all the polynomials in u with coefficients in K. It follows from equation (4) in §1 that

(1) $x \circ (y \circ v) = (x \circ y) \circ v$ for all $x, y \in K[u]$, if u and v commute.

In order to define the inverse of an element in $\mathcal{A}$ we prove

Lemma 5. *Let $u \in \mathcal{A}$. Then there exists an element $v \in \mathcal{A}$ such that $u \circ v = c$ and u, v commute if and only if u is not a zero divisor in $K[u]$. In this case, v is uniquely determined and belongs to $K[u]$.*

Proof. "$\Rightarrow$": Let $u \circ v = c$ and u, v commute. If $u \circ w = 0$, where $w \in K[u]$, we apply formula (1) and get $w = w \circ (u \circ v) = (u \circ w) \circ v = 0$.

"$\Leftarrow$": Conversely, assume that u is not a zero divisor of $K[u]$. Then the map $K[u] \to K[u], x \mapsto u \circ x$, is an injective endomorphism of the vector space $K[u]$. Hence this map is bijective and there exists a unique $x \in K[u]$ satisfying $u \circ x = c$. Obviously, u and x commute due to Corollary 1. Now let v be an arbitrary element of $\mathcal{A}$ such that $u \circ v = c$ and that u, v commute. Then (1) yields

$$x = x \circ (u \circ v) = (x \circ u) \circ v = v.$$

Thus x is the unique solution of this equation.   □

If $u \circ v = c$ and u, v commute we call v the *inverse* of u, in formulas $u^{-1} = v$. The equations $u^m \circ (u^{-1})^m = c$, $m \geq 1$, then yield that $(u^{-1})^{-1}$ and $(u^m)^{-1}(m \geq 2)$ exist and that

$$(u^{-1})^{-1} = u, \quad (u^{-1})^m = (u^m)^{-1}$$

hold. If u^{-1} exists, we define the powers u^{-m}, $m \geq 1$, by $u^{-m} := (u^{-1})^m$. (1) then implies

$$u^r \circ (u^s \circ v) = u^{r+s} \circ v, \quad r, s \text{ integers, provided that } u, v \text{ commute.}$$

In particular we have

$$u^r \circ u^s = u^{r+s} \in K[u] \quad \text{for all integers} \quad r, s.$$

There is another characterization of u^{-1} coming from the quadratic representation

$$P(u) = 2L^2(u) - L(u^2).$$

Theorem 12. *Let $u \in A$. Then the inverse u^{-1} exists if and only if $\det P(u) \neq 0$. In this case the following formulas are satisfied:*

$$u^{-1} = P^{-1}(u)u = P(u^{-1})u,$$
$$P(u)P(u^{-1}) = \text{Id},$$
$$L(u^{-1}) = L(u)P^{-1}(u) = P^{-1}(u)L(u).$$

Proof. "$\Rightarrow$": We assume that u^{-1} exists. Given an integer m the power u^m belongs to $K[u]$ and therefore all linear transformations $L(u^m)$ commute by pairs. Substituting $v = u^{-1}$ resp. $v = u^{-2}$ in equation (1) in §1, we therefore obtain

$$L(u) = P(u)L(u^{-1}) \quad \text{resp.} \quad 2L(u)L(u^{-1}) - P(u)L(u^{-2}) = \text{Id}.$$

If we substitute the first equation in the second one we conclude

$$P(u)P(u^{-1}) = \text{Id}.$$

Hence $\det P(u) \neq 0$.

"$\Leftarrow$": If u^{-1} does not exist, it follows from Lemma 5 that u is a divisor of zero in $K[u]$. Thus there is an element $x \in K[u]$ with $x \neq 0$ and $u \circ x = 0$. This yields $P(u)x = u \circ (u \circ x) = 0$, hence $\det P(u) = 0$.

If $\det P(u) \neq 0$ we get

$$P^{-1}(u)u = P(u^{-1})u = 2u^{-1} \circ (u^{-1} \circ u) - u^{-2} \circ u = u^{-1}.$$

$\square$

Suppose that $\det L(u) \neq 0$ for some $u \in \mathcal{A}$. Then u is not a zero divisor in $K[u]$ and thus u^{-1} exists. But the condition $\det L(u) \neq 0$ is not necessary for the existence of u^{-1}. If $\det L(u) \neq 0$, then $u \circ v = c$, $v \in \mathcal{A}$, implies $v = u^{-1}$ and in this case the equation $u \circ v = c$ has a unique solution in $\mathcal{A}$.

In addition to Theorem 3 we prove: If $x, y, x \circ y$ commute by pairs and if x^{-1} and y^{-1} exist, then we have

$$(2) \qquad\qquad (x \circ y)^{-1} = x^{-1} \circ y^{-1}.$$

Indeed, in view of $x^{-1} \in K[x]$, $y^{-1} \in K[y]$, it follows that $x^{-1} \circ y^{-1} \in K[x, y]$. We conclude from Theorem 3 that $x^{-1} \circ y^{-1}$ and $x \circ y$ commute. Since $K[x, y]$ is commutative and associative, we find $(x \circ y) \circ (x^{-1} \circ y^{-1}) = c$.

§6. Constructions of Jordan algebras

There is a standard method to obtain a Jordan algebra from an associative algebra. Let $\mathcal{B}$ be an associative algebra on the vector space X with the product $(u, v) \mapsto uv$. We define an algebra $\mathcal{A} = (X, \circ)$ by

$$u \circ v = \frac{1}{2}(uv + vu).$$

Obviously, $\mathcal{A}$ is a commutative algebra, and it is easy to see that $\mathcal{A}$ is a Jordan algebra. It follows from the definition that the powers of an element in $\mathcal{A}$ coincide with the powers in $\mathcal{B}$.

An easy computation shows that u and v commuting in $\mathcal{A}$, i.e $u \circ (v \circ z) = v \circ (u \circ z)$, for all z, is equivalent to $(uv - vu)z = z(uv - vu)$; equivalently $uv - vu$ belongs to the center of $\mathcal{B}$. Furthermore, we find that the quadratic representation is given by $P(u)v = uvu$. Suppose that $\mathcal{B}$ contains a unit element c. Then c is also the unit element of $\mathcal{A}$. We say that there exists the inverse in $\mathcal{B}$ of a given u if there is an element v in $\mathcal{B}$ such that $uv = vu = c$. Due to associativity, v is then uniquely determined by u. We show that the inverse of u in $\mathcal{B}$ exists if and only if the inverse of u in $\mathcal{A}$ exists, and that then the inverses are equal.

Suppose that $v \in \mathcal{B}$ satisfies $uv = vu = c$. Then u and v commute in $\mathcal{A}$ and therefore v is the inverse of u in $\mathcal{A}$.

Vice versa, let v be the inverse of u in $\mathcal{A}$, hence $v = P^{-1}(u)u$. Then we get

$$P(u)(c - uv) = u(c - uv)u = u^2 - u(P(u)v) = u^2 - u^2 = 0.$$

and $uv = c$. Analogously we get $vu = c$.

Therefore the inverse can unambiguously be denoted by u^{-1} in $\mathcal{A}$ and in $\mathcal{B}$.

Next we consider another method to construct Jordan algebras. We start with a vector space X over K, an element $c \in X$, a linear form $\lambda(x)$ and a symmetric bilinear form $\mu(x, y)$ on X. Let $\mathcal{A} = (X, \circ)$ be the algebra over K given by

$$x \circ y = \lambda(x)y + \lambda(y)x - \mu(x, y)c.$$

It is easy to verify that, whenever $\dim X \geq 2$, c is the unit element of $\mathcal{A}$ if and only if

(1) $$\lambda(x) = \mu(c, x), \quad \lambda(c) = 1.$$

Then the equality $x^2 = 2\lambda(x)x - \mu(x, x)c$ implies that $L(x^2)$ is a linear combination of $L(x)$ and Id. Hence $L(x^2)$ and $L(x)$ commute. *Hence, if (1), is satisfied, the product $(x, y) \mapsto x \circ y$ yields a Jordan algebra $\mathcal{A} = (X, \circ)$ with unit element c.*

We show that $\mu(x, y)$ is non-singular if $\mathcal{A}$ is semisimple.

Consider an element $x \in \mathcal{A}$ fulfilling $\mu(x, y) = 0$ for every $y \in \mathcal{A}$. Then $\lambda(x) = \mu(c, x) = 0$. Hence $x \circ y = \lambda(y)x$ and $x^2 = 0$ hold. We get $(x \circ y)^2 = \lambda^2(y)x^2 = 0$ and therefore $x \circ y$ is nilpotent for $y \in \mathcal{A}$. Theorem 6 shows that $x \in \operatorname{rad} \mathcal{A} = \{0\}$.

Conversely, assume that $\mu(x, y)$ is non-singular, and that $\operatorname{char} K$ does not divide $n := \dim X$. Given $u \in \mathcal{A}$ we define a linear form u^* by $u^*x = \mu(u, x)$. The linear transformation $x \mapsto \mu(u, x)v$ is denoted by vu^*; thus

$$\mu(u, x)v = (vu^*)x = (vx^*)u.$$

Given linear transformations U and V, denote their adjoint transformations (with respect to μ) by U^* and V^*. Then

$$(Uu)(Vv)^* = U(uv^*)V^*, \quad (uv^*)^* = vu^*.$$

immediately follow. The definition of $x \circ y$ yields

(2) $$L(x) = \lambda(x)\mathrm{Id} + xc^* - cx^*.$$

In view of $\operatorname{tr}(A^*) = \operatorname{tr}(A)$ and $\operatorname{tr}(vu^*) = \mu(u, v)$ we get $\operatorname{tr}L(x) = n\lambda(x)$. The bilinear form $\tau(u, v)$ associated with $\mathcal{A}$ (see §2) satisfies

(3) $$\frac{1}{n}\tau(x, y) = \frac{1}{n}\operatorname{tr}L(x \circ y) = \lambda(x \circ y) = 2\lambda(x)\lambda(y) - \mu(x, y).$$

Since $\mu(x, y)$ is non-singular, there exists a linear transformation V such that $\frac{1}{n}\tau(x, y) = \mu(Vx, y)$ and $V^* = V$. The last equation implies $V = 2cc^* - \mathrm{Id}$. As $V^2 = \mathrm{Id}$, the bilinear form $\tau(x, y)$ is non-singular.

Summarizing, we have proved that $\mathcal{A}$ is semisimple if and only if the bilinear form $\mu(x, y)$ is non-singular, provided that n is not a multiple of $\operatorname{char} K$.

Theorem 13. *Let X be a K–vector space of dimension n, with char K not dividing n, let $\tau(x,y)$ be a non-singular symmetric bilinear form on X and let $c \in X$ such that $\tau(c,c) = n$. Then the definitions*

$$x \circ y := \lambda(x)y + \lambda(y)x - \mu(x,y)c,$$

$$\lambda(x) := \frac{1}{n}\tau(c,x), \quad \mu(x,y) := 2\lambda(x)\lambda(y) - \frac{1}{n}\tau(x,y)$$

give rise to a semisimple Jordan algebra $\mathcal{A} = (X, \circ)$ with unit element c. The bilinear form $\mathrm{tr}L(x \circ y)$ associated with $\mathcal{A}$ is equal to $\tau(x,y)$.

Proof. (1) follows from $\tau(c,c) = n$. Hence $\mathcal{A}$ is a semisimple Jordan algebra. (3) implies $\mathrm{tr}L(x \circ y) = \tau(x,y)$. $\qquad\qquad\qquad\qquad\qquad\qquad\Box$

Now let $\mathcal{A} = (X, \circ)$ be the Jordan algebra described in Theorem 13. We show that $x, y \in \mathcal{A}$ commute if and only if x, y, c are linearly dependent. It is obvious that x, y commute whenever x, y, c are linearly dependent. Conversely, let x, y commute. The definition of the product yields

$$0 = x \circ (y \circ w) - y \circ (x \circ w)$$
$$= [\lambda(y)\lambda(w) - \mu(y,w)]x - [\lambda(x)\lambda(w) - \mu(x,w)]y$$
$$-[\lambda(y)\mu(x,w) - \lambda(x)\mu(y,w)]c.$$

If x, y, c are linearly independent, then we get $\lambda(y)\lambda(w) = \mu(y,w)$ and $\lambda(x)\lambda(w) = \mu(x,w)$ for every $w \in \mathcal{A}$, hence $y = \lambda(y)c$, $x = \lambda(x)c$. This is a contradiction. Hence x, y, c are linearly dependent.

What is the inverse of an element x in $\mathcal{A}$? Let us assume that x^{-1} exists. Since $x^{-1} \in K[x]$ and since $K[x]$ is spanned by c and x, we get $x^{-1} = \alpha x + \beta c$. Substituting this representation in $x \circ x^{-1} = c$, we obtain

$$(4) \qquad\qquad x^{-1} = \frac{1}{\mu(x,x)}\left[-x + 2\lambda(x)c\right], \quad \mu(x,x) \neq 0,$$

provided that c and x are linearly independent. But this equation also holds trivially for $x \in Kc$. Vice versa, let $\mu(x,x) \neq 0$. The right hand side of (4) defines an element u of $\mathcal{A}$ such that x, u commute and that $x \circ u = c$ holds. Therefore, x^{-1} exists if and only if $\mu(x,x) \neq 0$.

Remark. The Jordan algebra $\mathcal{A} = (X, \circ)$ can be obtained as a Jordan subalgebra of an associative algebra $\mathcal{B}$ by defining $x \circ y = \frac{1}{2}(xy + yx)$, where xy is the product in $\mathcal{B}$. In this case $\mathcal{B}$ can be chosen as the Clifford algebra over K of the restriction ν of μ to the subspace orthogonal to c. The dimension of this Clifford algebra is equal to 2^{n-1}. There exists an n–dimensional subspace X of $\mathcal{B}$ such that $X \circ X \subset X$. Then $\mathcal{A} = (X, \circ)$.

Notes

The notion of an algebra satisfying the axioms of a formally real Jordan algebra was introduced by P. Jordan in [14], [15]. The standard results presented in sections 1–4 (except Theorems 3 and 4) can be found in A.A. Albert's article [1] on the structure theory of Jordan algebras. For an introduction to Jordan algebras from the nonassociative algebra viewpoint see R.D. Schafer [23], [24].

The notion of commuting elements goes back to [15]. It was further investigated by N. Jacobson [6], [7], and B. Harris [3]. The inverse of an element in a Jordan algebra was first studied by N. Jacobson [8] in a different form; see also his article [10]. Here we work with a definition due to E. Artin, and follow his approach, which seems better suited for our purposes.

Editors' Notes

1. (i) The procedure of *polarizing identities*, i.e. replacing a variable u in an identity by $u + \lambda v$, and then "comparing coefficients" of the powers of λ, is used several times in this chapter. This procedure is fully explained in Braun and Koecher [BK], Kap. I, §2. Here, the reader can verify in every instance that the base field K has sufficiently many elements to allow the use of the procedure.

(ii) The notions of semisimplicity (via the trace form τ being non-singular) and simplicity introduced here are very useful for characteristic zero, which is the case of primary interest in this text. There are, however, some problems for positive characteristic; for instance there exist simple Jordan algebras (in the customary sense) which are not semisimple according to this definition.

(iii) The spaces $\mathcal{A}_\lambda(e)$, $\lambda = 0, 1/2, 1$, in Theorem 8 are usually called the *Peirce spaces* with respect to the idempotent e.

(iv) There is a shorter proof of Theorem 9: Since τ is non-singular, there exists a unique $c \in \mathcal{A}$ such that $\tau(c, x) = \operatorname{tr} L(x)$ for all x. Using $\tau(x, y) = \operatorname{tr} L(x \circ y)$ Lemma 4 shows that c is the unit element of $\mathcal{A}$. (See Braun and Koecher [BK], Kap. I, §6.) The proof presented here, though, contains useful additional information.

(v) A Jordan algebra with a unit element and with an arbitrary non-singular symmetric bilinear form $\mu(u, v)$ satisfying $\mu(u \circ v, w) = \mu(u, v \circ w)$ is semisimple. Hence Theorem 10 concerns semisimple Jordan algebras only. Moreover the element z in (ii) is uniquely determined and σ is non-singular if and only if z is invertible (cf. [BK], Kap. I, §6).

(vi) Likewise, Theorem 11 can be proved more quickly with the help of the trace form, using a theorem due to Dieudonne (cf. Schafer [23]): It follows from Theorem 6 that a semisimple Jordan algebra $\mathcal{A}$ contains no non-zero ideal with trivial multiplication. Now assume that $\mathcal{I}$ is an ideal of $\mathcal{A}$. Then Lemma 4 yields that $\mathcal{I}^{\perp} := \{y \in \mathcal{A};\ \tau(\mathcal{I}, y) = 0\}$ is also an ideal, and, furthermore, that $\mathcal{I} \circ \mathcal{I}^{\perp} = 0$. Therefore, $\mathcal{J} := \mathcal{I} \cap \mathcal{I}^{\perp}$ is an ideal with trivial multiplication, whence $\mathcal{J} = \{0\}$, and $\mathcal{A} = \mathcal{I} \oplus \mathcal{I}^{\perp}$. The decomposition theorem follows.

2. A fairly complete list of publications on Jordan algebras and related topics up to about 1964 may be found in [BK]. Since then, there has been a tremendous amount of work on Jordan algebras, in various directions. In the United States, the early work of Albert and Jacobson (see the important and influential monograph [Jac2]) was continued by Jacobson and his school. In Russia, a group headed by Shirshov worked on certain classes of nonassociative algebras (including Jordan algebras), using powerful combinatorial tools; see [ZSSS]. A culmination point was Zelmanov's structure theory on infinite dimensional Jordan algebras. The survey article [McC2] by McCrimmon contains a very well-written outline of the development up to the mid-1980s. Infinite dimensional Jordan (operator) algebras are also being investigated in functional analysis; we only mention the monograph [H-OS] by Hanche-Olsen and Størmer, and several of the contributions to [KMP].

The above outline is not complete in any way. Recent developments (including extensive references) are described in the conference proceedings [KMP] and [F-L]. In addition, there is now an internet address where preprints of recent articles on Jordan structures and related topics may be found. The URL is

$$\texttt{http://math1.uibk.ac.at/jordan/}$$

Koecher's approach to Jordan algebras is motivated by their applications in finite dimensional analysis and number theory, and therefore his investigations are mainly restricted to finite dimensional real or complex Jordan algebras. It is worth pointing out once more that the monograph [FKo] by Faraut and Korányi is an excellent contemporary source on Jordan algebras in analysis.

Emphasis on properties of the inverse is characteristic for Koecher's approach; see also Braun and Koecher [BK]. Indeed one may say that Jordan algebras are distinguished by "good" properties of the inverse. Springer [Spr] later built a theory of finite dimensional Jordan algebras using properties of the inverse as axioms; see also McCrimmon [McC1].

Chapter IV. Real and Complex Jordan Algebras

Throughout this chapter let $\mathcal{A}$ be a Jordan algebra with unit element c and of finite dimension n over the field K. Moreover, K is either the field $\mathbf{R}$ of real numbers, or the field $\mathbf{C}$ of complex numbers. We restrict our investigations to this case because we will use the differential operators introduced in Chapter I, §6. Since these operators are also well-defined for fields of characteristic zero or greater than n, almost all the results of this chapter also hold in the general case.

§1. The quadratic representation

In §1 and §5 of Chapter III we considered the so-called quadratic representation $x \mapsto P(u)x = 2u \circ (u \circ x) - u^2 \circ x$, of the Jordan algebra $\mathcal{A}$,

$$P(u) = 2L^2(u) - L(u^2).$$

Theorem III.12 states that the inverse u^{-1} of an element $u \in \mathcal{A}$ exists if and only if $\det P(u) \neq 0$. In this case $u^{-1} = P^{-1}(u)u$. We apply $\frac{\partial}{\partial x}$ (see Chapter I, §6 and Chapter II, §4) to $x \circ x^{-1} = c$ and $x^2 \circ x^{-1} = x$, and obtain

$$L(x)\frac{\partial x^{-1}}{\partial x} + L(x^{-1}) = 0, \quad L(x^2)\frac{\partial x^{-1}}{\partial x} + 2L(x^{-1})L(x) = Id.$$

Multiplying the first equation by $2L(x)$ and subtracting the second formula yields (recalling that $L(x)$ and $L(x^{-1})$ commute)

(1) $$\frac{\partial x^{-1}}{\partial x} = -P^{-1}(x) \quad \text{for} \quad \det P(x) \neq 0.$$

Theorem 1. *Let $x, y \in \mathcal{A}$. Then the "fundamental formula"*

$$P(P(x)y) = P(x)P(y)P(x)$$

holds. The inverse of $P(x)y$ exists if and only if x^{-1} and y^{-1} exist. Then

$$(P(x)y)^{-1} = P^{-1}(x)y^{-1}.$$

follows.

Proof. We start with two elements $x, y \in \mathcal{A}$ for which x^{-1} and y^{-1} exist. It follows that

$$(2) \quad \Delta_x^{y^{-1}} P(x)y = 2y^{-1} \circ (x \circ y) + 2x \circ (y^{-1} \circ y) - 2(y^{-1} \circ x) \circ y = 2x.$$

Theorem III.12 states that $L(x^{-1}) = L(x)P^{-1}(x)$ and hence

$$x^{-1} \circ P(x)y = x \circ y.$$

We apply $\Delta_x^{y^{-1}}$ to this formula. Then (1) and (2) lead to

$$P(x)y \circ P^{-1}(x)y^{-1} = c.$$

Given x, y such that $\det L(P(x)y) \neq 0$ it follows (cf. Theorem III.12) that

$$(P(x)y)^{-1} = P^{-1}(x)y^{-1}.$$

Thus (1) yields

$$\frac{\partial (P(x)y)^{-1}}{\partial y} = \frac{\partial (P(x)y)^{-1}}{\partial P(x)y} \cdot \frac{\partial P(x)y}{\partial y} = -P^{-1}(P(x)y)P(x),$$

$$\frac{\partial (P^{-1}(x)y^{-1})}{\partial y} = P^{-1}(x)\frac{\partial y^{-1}}{\partial y} = -P^{-1}(x)P^{-1}(y).$$

Therefore, the fundamental formula has been proved for all $(x, y) \in \mathcal{A} \times \mathcal{A}$ such that $\det P(x) \neq 0$, $\det P(y) \neq 0$ and $\det L(P(x)y) \neq 0$. Since these points form a non-empty open subset of $\mathcal{A} \times \mathcal{A}$, and since $P(u)$ is a polynomial in u, the fundamental formula holds in $\mathcal{A} \times \mathcal{A}$. From this formula it follows that $(P(x)y)^{-1}$ exists if and only if x^{-1} and y^{-1} exist. Using the representation $u^{-1} = P^{-1}(u)u$ and the fundamental formula, the last assertion follows.   □

Substituting $y = c, x, x^2, \dots$ in the fundamental formula, we obtain a generalization of Theorem III.4:

$$P^m(x) = P(x^m), \quad m \geq 1.$$

There is, in turn, a generalization of this formula:

Theorem 2. *Let $x, y, x \circ y$ commute pairwise. Then one has*

$$P(x \circ y) = P(x)P(y).$$

Proof. Theorem III.3 states that $u^k, v^l, u \circ v$ commute if u, v are linear combinations of $x^r \circ y^s$ for $r, s \geq 0$. We therefore get $P(u)v^2 = u^2 \circ v^2 = (u \circ v)^2$. The fundamental formula then shows (cf. Theorem III.4)

$$P^2(u \circ v) = P(u^2 \circ v^2) = P(u)P(v^2)P(u) = (P(u)P(v))^2.$$

This yields $0 = [P(u \circ v) - P(u)P(v)][P(u \circ v) + P(u)P(v)]$. Substituting $u = c + \lambda x$, $v = c + \lambda y$ leads to

$$\det[P((c + \lambda x) \circ (c + \lambda y)) + P(c + \lambda x)P(c + \lambda y)] \neq 0,$$

except for at most finitely many $\lambda \in K$. Therefore

$$P((c + \lambda x) \circ (c + \lambda y)) = P(c + \lambda x)P(c + \lambda y)$$

holds except for those finitely many λ. Since both sides are polynomials in λ, this formula holds without exception. Comparing the coefficients of λ^4 proves the theorem. $\square$

A Jordan algebra may be characterized by the fundamental formula in the following way. For each $u \in X$, let $Q(u)$ be a linear transformation on X and let c be an element of X such that:

(a) The map $X \to Hom(X, X)$, $u \mapsto Q(u)$, is a homogeneous polynomial of degree 2.

(b) $Q(c) = \text{Id}$.

(c) $Q(c + u) - \text{Id} - Q(u) = 0$ implies $u = 0$.

(d) $Q(Q(u)v) = Q(u)Q(v)Q(u)$ for every $u, v \in X$.

We know that in a Jordan algebra $\mathcal{A}$ with unit element c the quadratic representation $Q(u) = P(u)$ possesses these properties. Vice versa we obtain

Theorem 3. *Let $Q(u)$ be a linear transformation on the vector space X satisfying the properties (a)–(d). Then the composition*

$$u \circ v := \frac{1}{2}(Q(u + v)c - Q(u)c - Q(v)c)$$

defines a Jordan algebra $\mathcal{A} = (X, \circ)$ with unit element c and $Q(u)$ as its quadratic representation.

Proof. It follows from (a) that for each pair (u, v) there is a linear transformation $Q(u, v)$ such that

$$Q(u + v) = Q(u) + 2Q(u, v) + Q(v).$$

$Q(u, v)$ is bilinear and symmetric in u and v. Let

$$L(u) = Q(u, c) = Q(c, u).$$

Then (c) shows that $L(u) = L(v)$ implies $u = v$. Moreover,

$$Q(c + u) = \text{Id} + 2L(u) + Q(u)$$

holds.

Now we define a bilinear composition

$$u \circ v := L(u)v.$$

We substitute $u = c + \lambda x$, $v = c + \mu y$, $\lambda, \mu \in K$, in (d) and compute the terms which are of degree one, resp. bilinear, in λ and μ, on both sides:

$$Q(Q(u)v) = Q(c + \mu y + 2\lambda x \circ c + 2\lambda \mu x \circ y + \dots)$$
$$= \mathrm{Id} + 2\mu L(y) + 4\lambda L(x \circ c) + 4\lambda \mu L(x \circ y) + 4\lambda \mu Q(y, x \circ c) + \dots ,$$
$$Q(u)Q(v)Q(u) = (\mathrm{Id} + 2\lambda L(x) + \dots)(\mathrm{Id} + 2\mu L(y) + \dots)(\mathrm{Id} + 2\lambda L(x) + \dots)$$
$$= \mathrm{Id} + 2\mu L(y) + 4\lambda L(x) + 4\lambda \mu L(x)L(y) + 4\lambda \mu L(y)L(x) + \dots .$$

A comparison yields $L(x) = L(x \circ c)$ and therefore $x = x \circ c$ for every $x \in \mathcal{A}$. Moreover, we obtain

(3) $$Q(y, x) = L(x)L(y) + L(y)L(x) - L(x \circ y).$$

Since the left hand side is symmetric in x, y, we have $x \circ y = y \circ x$, i.e. the composition is commutative. Substituting $x = y$ we get

$$Q(x) = Q(x, x) = 2L^2(x) - L(x^2).$$

This shows $Q(x)c = x^2$. Substituting $v = c$ in (d) leads to $Q(u^2) = Q^2(u)$, and Theorem III.4 implies that $\mathcal{A}$ is a Jordan algebra with unit element c. It follows from (3) that $x \circ y = Q(x, y)c$ and that Q is its quadratic representation. $\qquad\square$

§2. Mutations

Given an element f in the Jordan algebra $\mathcal{A} = (X, \circ)$ with unit element c we define a new composition $u \perp v$ on the vector space X by

$$u \perp v := u \circ (v \circ f) + v \circ (u \circ f) - (u \circ v) \circ f.$$

It is obvious that $u \perp v$ is a bilinear and commutative composition. We denote this new algebra $(X, \perp)$ by $\mathcal{A}_f$ and call $\mathcal{A}_f$ a *mutation* of $\mathcal{A}$. If we describe the composition of $\mathcal{A}_f$ by the linear transformation $L_f(u)$,

$$u \perp v =: L_f(u)v,$$

we immediately obtain

(1) $$L_f(u) = L(u)L(f) - L(f)L(u) + L(u \circ f).$$

Moreover, we define

$$P_f(u) := 2L_f^2(u) - L_f(u \perp u).$$

Obviously $u \perp u = P(u)f$ holds. This shows

$$\frac{\partial P(u)f}{\partial u} = 2L_f(u).$$

Let us now discuss under which conditions A_f possesses a unit element c_f. Substituting $v = c$, $u = c_f$ in the equation defining $u \perp v$, we get $c_f \circ f = c$. Then $v = c_f$ yields $c_f \circ (u \circ f) = (u \circ c_f) \circ f$. Hence f and c_f commute, and it follows from III, §5 that $c_f = f^{-1}$. Conversely, if f^{-1} exists then f^{-1} is the unit element of the mutation A_f.

Theorem 4. *Let A be a Jordan algebra with unit element c and $f \in A$. Then the mutation A_f is a Jordan algebra, and its quadratic representation is given by*

$$P_f(u) = P(u)P(f).$$

Proof. Since the product in A_f depends linearly on f it is sufficient to prove

$$L_f(u \perp u)L_f(u) = L_f(u)L_f(u \perp u) \quad \text{and} \quad P_f(u) = P(u)P(f)$$

for invertible elements $f \in A$. We therefore assume $\det P(f) \neq 0$ and define

$$Q(u) := P(u)P(f), \quad c_f := f^{-1}.$$

Since $Q(u)$ is a homogeneous polynomial of degree 2, we get

$$Q(c_f + u) = \mathrm{Id} + 2M(u) + Q(u)$$

with $M(u)$ linear in u. In order to calculate $M(u)$, we use

$$P(u + v) = P(u) + 2P(u, v) + P(v),$$

with $P(u, v) = L(u)L(v) + L(v)L(u) - L(u \circ v)$, and find,

(2) $$M(u) = P(u, f^{-1})P(f).$$

In view of $M(u)c_f = P(u, f^{-1})f = u$, the map $A \to A, u \mapsto M(u)$, is injective. The fundamental formula for A in Theorem 1 yields $Q(Q(u)v) = Q(u)Q(v)Q(u)$. According to Theorem 3, the product $u * v := M(u)v$ defines a Jordan algebra on the vector space A, with unit element c_f. This composition can be written as

(3) $$u * v = \frac{1}{2}[Q(u + v)c_f - Q(u)c_f - Q(v)c_f].$$

By virtue of $Q(u)c_f = P(u)f$ we immediately get $u * v = P(u, v)f = u \perp v$, and thus $M(u) = L_f(u)$. □

Corollary 1. *Given u, f in a Jordan algebra A with unit element c one has*
(a) $L_f(u) = P(u, f^{-1})P(f)$ if $\det P(f) \neq 0$.
(b) $L_f(u)P(u) = P(u)L_u(f)$.

Proof. (a) The claim follows from (1) and $M(u) = L_f(u)$.

(b) It is sufficient to prove this identity in the case $\det P(f) \neq 0$ as well as $\det P(u) \neq 0$. We consider the mutation A_f. The inverse of u in A_f is given by $u^{\sharp} = P_f^{-1}(u)u = P^{-1}(f)u^{-1}$. Now Theorem III.12 and part (a) imply

$$P_f^{-1}(u)L_f(u) = L_f(u^{\sharp}) = P(u^{\sharp}, f^{-1})P(f) = P(P^{-1}(f)u^{-1}, f^{-1})P(f),$$

hence

$$P^{-1}(u)L_f(u) = P(f)P(P^{-1}(f)u^{-1}, f^{-1})P(f).$$

The fundamental formula in Theorem 1 leads to

$$P(a)\, P(u,v)\, P(a) \;=\; P(P(a)u,\, P(a)v)$$

and therefore we obtain $P^{-1}(u)L_f(u) = P(u^{-1}, f) = L_u(f)P^{-1}(u)$. This proves (b).   □

Now let us consider the bilinear form

$$\tau_f(u,v) = \operatorname{tr} L_f(u \perp v)$$

associated with the mutation A_f of A. By virtue of $\operatorname{tr}L_f(u) = \operatorname{tr}L(u \circ f) = \tau(u,f)$ we get $\tau_f(u,v) = \tau(u \perp v, f) = \tau(u \circ (v \circ f), f) + \tau(v \circ (u \circ f), f) - \tau((u \circ v) \circ f, f) = 2\tau(v, f \circ (f \circ u)) - \tau(v, f^2 \circ u)$ or

$$\tau_f(u,v) = \tau(P(f)u, v).$$

Let A be semisimple; then this identity shows that A_f is semisimple if and only if f^{-1} exists.

§3. A generalization of the fundamental formula

Given a Jordan algebra A we define

$$\pi(u) = \det P(u),$$

where $P(u)$ is the quadratic representation of A. The fundamental formula in Theorem 1 implies that

$$\pi(P(u)v) = \pi^2(u)\pi(v) \quad \text{for} \quad u, v \in A.$$

As to recovering P from π, one has:

Lemma 1. *If* $\pi(x) \neq 0$ *then*

$$\Delta_x^u \log \pi(x) = 2\tau(x^{-1}, u), \quad and \quad \Delta_x^v \Delta_x^u \log \pi(x) = -2\tau(P^{-1}(x)v, u).$$

Proof. For $\lambda \in K$ we consider $\det P(x + \lambda u)$. In view of $\det P(x) \neq 0$ we get

$$\det P(x + \lambda u) = \det \left(P(x) + 2\lambda P(x, u) + \lambda^2 P(u) \right)$$
$$= \det P(x) \cdot \det \left(Id + 2\lambda P(x, u) P^{-1}(x) + O(\lambda^2) \right)$$
$$= \det P(x) \cdot \left(1 + 2\lambda \operatorname{tr} \left(P(x, u) P^{-1}(x) \right) + O(\lambda^2) \right).$$

Now we apply Corollary 1(a) and obtain

$$\pi(x + \lambda u) = \pi(x)(1 + 2\lambda \operatorname{tr} L_{x^{-1}}(u) + O(\lambda^2))$$
$$= \pi(x)(1 + 2\lambda \tau(x^{-1}, u) + O(\lambda^2)).$$

Thus the first assertion has been proved. The second assertion follows from $\frac{\partial x^{-1}}{\partial x} = -P^{-1}(x)$.

$\square$

In particular, a semisimple Jordan algebra satisfies

$$(1) \qquad \left(\frac{\partial \log \pi(x)}{\partial x} \right)^* = 2x^{-1}, \quad \frac{\partial}{\partial x} \left(\frac{\partial \log \pi(x)}{\partial x} \right)^* = -2P^{-1}(x),$$

where the gradient is taken with respect to the bilinear form $\tau(u, v)$.

To generalize the fundamental formula, we consider the set $\Gamma(\mathcal{A})$ of all those non-singular linear transformations W on the vector space $\mathcal{A}$ for which there exists a linear transformation V such that

$$(\dagger) \qquad\qquad P(Wu) = WP(u)V \quad \text{for all} \quad u \in \mathcal{A}.$$

Obviously V is uniquely determined by W; for instance one has $V = W^{-1} P(Wc)$; hence we can define

$$W^* := V,$$

and W^* is non-singular. It is easy to see that $\Gamma(\mathcal{A})$ is a group, which is called the *structure group* of $\mathcal{A}$. The fundamental formula in Theorem 1 shows that $P(x)$ belongs to $\Gamma(\mathcal{A})$, and that $P^*(x) = P(x)$ holds, provided that $\det P(x) \neq 0$. This immediately yields that $W \in \Gamma(\mathcal{A})$ implies $W^* \in \Gamma(\mathcal{A})$.

Clearly a pair $(W, V) \in GL(\mathcal{A}) \times GL(\mathcal{A})$ satisfies equation $(\dagger)$ if and only if the matrix coefficients of W and V (with respect to some basis) satisfy a finite system of polynomial equations. From this and the equality $W^* = W^{-1} P(Wc)$ it follows that $\Gamma(\mathcal{A})$ is closed in $GL(\mathcal{A}) \times GL(\mathcal{A})$; hence a Lie group.

As usual we call an algebra isomorphism from $\mathcal{A}$ onto $\mathcal{A}$ an *automorphism of the Jordan algebra*. A non-singular linear transformation W on $\mathcal{A}$ turns out to be an automorphism if and only if $W(x \circ y) = Wx \circ Wy$ or $WL(x) = L(Wx)W$.

Lemma 2. *The group $\mathrm{Aut}(\mathcal{A})$ of all automorphisms of a Jordan algebra $\mathcal{A}$ is equal to the set of all $W \in \Gamma(\mathcal{A})$ which satisfy $Wc = c$.*

Proof. Let W be an automorphism. Then the definition yields $WP(u) = P(Wu)W$. Hence one obtains $W \in \Gamma(\mathcal{A})$, with $W^* = W^{-1}$.

Let $W \in \Gamma(\mathcal{A})$ satisfy $Wc = c$. Substituting $u = c$ in the definition of $\Gamma(\mathcal{A})$ we get $W^* = W^{-1}$, and this leads to

$$(Wu)^2 = P(Wu)c = WP(u)W^{-1}c = Wu^2.$$

By polarization $Wx \circ Wy = W(x \circ y)$ follows. Thus W is an automorphism.
□

Now we show that for a semisimple Jordan algebra $\mathcal{A}$ the transformation W^* is equal to the adjoint transformation of W with respect to the bilinear form $\tau(x, y) = \mathrm{tr}L(x \circ y)$.

Theorem 5. *Let $\mathcal{A}$ be a semisimple Jordan algebra. A non-singular linear transformation W belongs to the structure group $\Gamma(\mathcal{A})$ if and only if $\pi(Wx) = \gamma \cdot \pi(x)$ for some γ. In this case, W^* is the adjoint transformation of W with respect to $\tau(x, y)$.*

Proof. Obviously we get $\pi(Wx) = \gamma \cdot \pi(x)$ for $W \in \Gamma(\mathcal{A})$. Conversely, start with a non-singular linear transformation W and assume $\pi(Wx) = \gamma \cdot \pi(x)$. Using the second identity of Lemma 1 yields

$$P(Wx) = WP(x)W^*,$$

where W^* is the adjoint transformation with respect to $\tau(x, y)$.
□

Given a semisimple Jordan algebra $\mathcal{A}$, there is another description of $\Gamma(\mathcal{A})$. We consider $\mathrm{tr}\,(AP(u))$ for a given linear transformation A on $\mathcal{A}$. This is a quadratic form in u, and thus there exists a self-adjoint linear transformation $T(A)$ such that

$$\mathrm{tr}\,(AP(u)) = \tau(T(A)u, u).$$

Obviously $T(A)$ is a linear map and homogeneous in A. The identity

$$\tau((W^*T(A)W - T(W^*AW))u, u) = \mathrm{tr}[A(P(Wu) - WP(u)W^*)]$$

immediately shows that a non-singular transformation W belongs to $\Gamma(\mathcal{A})$ if and only if

$$T(W^*AW) = W^*T(A)W \quad \text{for every} \quad A.$$

We substitute $W = P(c + u)$ and compare the linear terms in u. Using the Jordan product $A \circ B = \frac{1}{2}(AB + BA)$ for linear transformations we get

$$T(L(u) \circ A) = L(u) \circ T(A)$$

for $u \in \mathcal{A}$ and an arbitrary linear transformation A.

With regard to Theorem 5 we call two Jordan algebras $\mathcal{A}$ and $\tilde{\mathcal{A}}$ *similar* if there exists a non-singular linear transformation $W : \mathcal{A} \to \tilde{\mathcal{A}}$ such that $\tilde{\pi}(Wu) = \gamma \cdot \pi(u)$ holds for every $u \in \mathcal{A}$, where $\tilde{\pi}(u)$ resp. $\pi(u)$ are the determinants of the quadratic representations $\tilde{P}(u)$ in $\tilde{\mathcal{A}}$ resp. $P(u)$ in $\mathcal{A}$.

Theorem 6. *Let $\mathcal{A}$ and $\tilde{\mathcal{A}}$ be semisimple Jordan algebras. Then $\tilde{\mathcal{A}}$ is similar to $\mathcal{A}$ if and only if $\tilde{\mathcal{A}}$ is isomorphic to a mutation $\mathcal{A}_f$ of $\mathcal{A}$.*

Corollary 2. *We have $\tilde{\mathcal{A}} \cong \mathcal{A}_f$ for some invertible $f \in \mathcal{A}$ if and only if there exists an invertible linear transformation $W : \mathcal{A} \to \tilde{\mathcal{A}}$ such that $\tilde{\pi}(Wu) = \gamma \cdot \pi(u)$ for some $\gamma \neq 0$.*

Proof. If $\tilde{\mathcal{A}}$ is isomorphic to a mutation $\mathcal{A}_f$ of $\mathcal{A}$ then the claim follows immediately from Theorem 4, since $\mathcal{A}$ has a unit element due to Theorem III.9. Conversely, we apply Lemma 1 to $\tilde{\pi}(Wx) = \gamma \cdot \pi(x)$, hence

$$\tau(P^{-1}(x)v, u) = \tilde{\tau}(\tilde{P}^{-1}(Wx)Wv, Wu).$$

Here τ, $\tilde{\tau}$ denote the associated bilinear forms on $\mathcal{A}$ and $\tilde{\mathcal{A}}$. Since τ and $\tilde{\tau}$ are non-singular, there is a non-singular self-adjoint transformation T such that $\tilde{\tau}(Wv, Wu) = \tau(Tv, u) = \tau(v, Tu)$. Thus we obtain

$$(2) \qquad \tilde{P}(Wx) = WP(x)W^\sharp, \quad W^\sharp = TW^{-1}.$$

Let $\tilde{c}$ resp. c and $\tilde{\circ}$ resp. $\circ$ be the unit elements and product in $\tilde{\mathcal{A}}$ resp. $\mathcal{A}$. Substituting $x = W^{-1}\tilde{c}$ in (2) we see that $W^{-1}\tilde{c}$ is invertible in $\mathcal{A}$, hence

$$f^{-1} = W^{-1}\tilde{c}$$

for some $f \in \mathcal{A}$. Thus we get $P(f) = T$, hence $f = P(f)f^{-1} = W^\sharp \tilde{c}$. Let $\mathcal{A}_f$ be the mutation of $\mathcal{A}$ with respect to f and denote by $\perp$ the composition in $\mathcal{A}_f$. From (2) it follows that

$$Wx \,\tilde{\circ}\, Wx = \tilde{P}(Wx)\tilde{c} = WP(x)W^\sharp \tilde{c} = WP(x)f = W(x \perp x),$$

therefore $Wx \,\tilde{\circ}\, Wy = W(x \perp y)$. Hence $W : \mathcal{A}_f \to \tilde{\mathcal{A}}$ is an isomorphism. $\square$

Now we may ask under which conditions two mutations $\mathcal{A}_f$ and $\mathcal{A}_g$ for invertible elements $f, g \in \mathcal{A}$ are isomorphic. This is equivalent to the existence of a non-singular linear transformation W such that $W(u \perp_f v) = Wu \perp_g Wv$ or to

$$(3) \qquad\qquad WL_f(u) = L_g(Wu)W.$$

This yields

$$(4) \qquad\qquad WP_f(u) = P_g(Wu)W.$$

In view of $P_f(u) = P(u)P(f)$ due to Theorem 4 we get $W \in \Gamma(\mathcal{A})$ and $W^*P(g)W = P(f)$. We apply $\partial/\partial x$ to $P(Wx)g = WP(x)W^*g$, then $\frac{\partial P(x)y}{\partial x} = 2L_y(x)$ immediately yields

$$L_g(Wx)W = WL_{W^{\cdot}g}(x).$$

Together with (3) it follows that $L_f(x) = L_{W^*g}(x)$, hence $f = W^*g$. Therefore $\mathcal{A}_f \cong \mathcal{A}_g$ implies $f = W^*g$ for some $W \in \Gamma(\mathcal{A})$. Conversely, let $f = W^*g$, where $W \in \Gamma(\mathcal{A})$. Then we get (4) and by $u = f^{-1}$ also $g^{-1} = Wf^{-1}$. We substitute $u = f^{-1} + x$ in (4) and equation (3) arises from comparing the linear terms in x. Hence $\mathcal{A}_f \cong \mathcal{A}_g$ follows.

Summarizing we have

Theorem 7. *Two semisimple mutations $\mathcal{A}_f$ and $\mathcal{A}_g$ of $\mathcal{A}$ are isomorphic if and only if there is a transformation W in the structure group $\Gamma(\mathcal{A})$ such that $f = W^*g$. In this case, the isomorphism is given by $W : \mathcal{A}_f \to \mathcal{A}_g$.*

What happens if we continue this process of constructing mutations? For this we have to consider $(\mathcal{A}_f)_g$. Using the fact that a Jordan algebra is uniquely determined by its quadratic representation and its unit element because of Theorem 3, we get

$$(\mathcal{A}_f)_g = \mathcal{A}_{P(f)g}.$$

Now Theorem 7 shows that $(\mathcal{A}_f)_g$ is always isomorphic to $\mathcal{A}_g$, whenever f, g are invertible.

§4. The exponential

Let us consider the *exponential function*

$$e(u) = \sum_{m=0}^{\infty} \frac{1}{m!} u^m.$$

Then $e(0) = c$, and in view of $u^m = L^m(u)c$ we get $e(u) = e^{L(u)}c$. Hence it follows that the power series for $e(u)$ is absolutely and locally uniformly convergent on $\mathcal{A}$. Therefore $e(u)$ is real-analytic on $\mathcal{A}$. Since the Jacobian in $u = 0$ is Id, there exists a neighborhood of 0 in $\mathcal{A}$ such that the map $u \mapsto e(u)$ is a bijection onto a neighborhood of c in $\mathcal{A}$.

Lemma 3. *Suppose that x, y and $x \circ y$ commute by pairs, then*

$$e(x) \circ e(y) = e(x + y), \quad and \quad e(x), \ e(y), \ e(x + y) \quad commute \ pairwise.$$

Proof. Let $K[x, y]$ denote the vector space spanned by $x^r \circ y^s$ for $r, s \geq 0$. From Theorem III.3 it follows that $K[x, y]$ is an associative subalgebra of $\mathcal{A}$ and that the elements of $K[x, y]$ commute pairwise. Now the standard proof of the functional equation can be carried over to this situation.   □

In particular this equation holds for $x, y \in K[u]$, $u \in \mathcal{A}$. Theorem 2 and Lemma 3 show that

(1) $P(e(x))P(e(y)) = P(e(x + y))$ if $x, y, x \circ y$ commute pairwise.

Lemma 4. *Given $u \in \mathcal{A}$ one has*

$$P(e(u)) = e^{2L(u)}.$$

Proof. Substituting $x = \alpha u$, $y = \beta u$ in (1) yields

(2) $P(e((\alpha + \beta)u)) = P(e(\alpha u)) \, P(e(\beta u)).$

Let

$$P(e(u)) = \sum_{m=0}^{\infty} \frac{1}{m!} A_m(u)$$

be the expansion of $P(e(u))$ in terms of homogeneous polynomials $A_m(u)$ of degree m. We substitute this expansion in (2) and compare the coefficients of $\alpha \beta^{m-1}$:

$$\frac{m}{m!} A_m(u) = \frac{1}{(m-1)!} A_1(u) A_{m-1}(u).$$

Therefore we get $A_m(u) = A_1^m(u)$ by induction. In order to calculate $A_1(u)$ we consider

$$P(e(u)) = P(c + u + \ldots) = \mathrm{Id} + 2L(u) + \ldots$$

This shows $A_1(u) = 2L(u)$.   □

Lemma 4 immediately yields

(3) $$\pi(e(u)) = \det P(e(u)) = e^{2\mathrm{tr}L(u)} = e^{2\tau(c,u)}.$$

Now let us consider the structure group $\Gamma(\mathcal{A})$ in a neighborhood of the identity.

Lemma 5. *There exists a neighborhood U of the identity in the structure group $\Gamma(\mathcal{A})$ such that any $W \in U$ can be written in the form*

$$W = e^{L(x)}V = P(y)V,$$

where V is an automorphism of $\mathcal{A}$ belonging to a neighborhood of Id, and $y = e\left(\frac{1}{2}x\right)$ holds. There is a neighborhood $\tilde{U}$ of c such that y can be chosen in $\tilde{U}$, and y and V are then uniquely determined by W.

Proof. If U is sufficiently small, we can find an element $x \in \mathcal{A}$ such that $Wc = e(x)$ and there is a neighborhood of c such that x is unique in this neighborhood. Put $V = P^{-1}\left(e(\frac{1}{2}x)\right)W$. Then we get $Vc = P^{-1}(e\left(\frac{1}{2}x\right))e(x) = c$. Since V belongs to $\Gamma(\mathcal{A})$, the transformation V is an automorphism of $\mathcal{A}$ due to Lemma 2. The last equation follows from Lemma 4. $\qquad\square$

This lemma implies that for x and y in a suitable neighborhood of c there exist an element $u = u(x,y)$ and an automorphism $V = V(x,y)$ such that

$$P(x)P(y) = P(u)V.$$

Applying this formula to c, we see that u is a solution of $u^2 = P(x)y^2$. Conversely, if u is a solution of $u^2 = P(x)y^2$ then $P^{-1}(u)P(x)P(y)$ is an automorphism of $\mathcal{A}$.

In view of $e^2(x) = e(2x)$, and since $x \mapsto e(x)$ maps a neighborhood of 0 onto a neighborhood of c, we conclude that $u \mapsto u^2$ maps a neighborhood U_1 of c bijectively onto a neighborhood U_2 of c. In U_2 we therefore get a uniquely determined square root $u^{1/2}$ of u such that $(u^{1/2})^2 = u$ for all $u \in U_2$, and $(u^2)^{1/2} = u$ for all $u \in U_1$. By the usual argument we have

$$(c+x)^{1/2} = c + \frac{1}{2}x - \frac{1}{2}x^2 + \ldots = \sum_{m=0}^{\infty} \binom{1/2}{m} x^m$$

for small x.

By virtue of $P^2(u^{1/2}) = P(u)$, the inverse of $u^{1/2}$ exists if the inverse of u exists. We write $u^{-1/2}$ for $(u^{1/2})^{-1}$.

Now we define for u, v in a neighborhood of c the transformation

$$V(u,v) := P\left((P(u)v^2)^{-1/2}\right) P(u) P(v).$$

Lemma 6. *Given u, v in a suitable neighborhood of c the linear transformation $V(u, v)$ is an automorphism of A and one has*

$$V(c + \lambda x, \ c + \lambda y) = \mathrm{Id} - 2\lambda^2 (L(x)L(y) - L(y)L(x)) + O(\lambda^3).$$

Proof. Obviously $V(u, v)$ belongs to $\Gamma(A)$. On the other hand we obtain, abbreviating $w = (P(u)v^2)^{1/2}$,

$$V(u, v)c = P(w^{-1})P(u)v^2 = P(w^{-1})w^2 = c.$$

Thus Lemma 2 shows that $V(u, v)$ is an automorphism of A. For the remaining equation we put $u = c + \lambda x$, $v = c + \lambda y$. Then we have (using the power series for $(c + x)^{-1/2}$)

$$P(u)v^2 = c + 2\lambda(x + y) + \lambda^2(x^2 + 4x \circ y + y^2) + O(\lambda^3),$$
$$(P(u)v^2)^{-1/2} = c - \lambda(x + y) + \lambda^2(x^2 + x \circ y + y^2) + O(\lambda^3).$$

Substituting this formula in $V(u, v)$ and using $P(a + b) = P(a) + 2P(a, b) + P(b)$ we get the desired result. $\qquad\qquad\square$

Note that indeed

$$P(x)P(y) = P\left((P(u)v^2)^{1/2}\right) V(u, v)$$

for all u, v near c. Thus $V(u, v)$ is the automorphism introduced after Lemma 5.

§5. The associated Lie algebra

Let Γ be a Lie group of linear transformations. A linear transformation T is called an *infinitesimal transformation* of Γ if there exists a twice continuously differentiable curve $W = W(\tau) \in \Gamma$, $|\tau| \leq \varepsilon$, such that $W(0) = \mathrm{Id}$, $\dot{W}(0) = T$. Denote the set of all infinitesimal transformations of Γ by $\mathcal{L}(\Gamma)$. It is well-known that $\mathcal{L}(\Gamma)$ is a vector space over K. We will investigate the space $\mathcal{L}(\Gamma)$ for

$$\Gamma = \Gamma(A) = \{W; \ P(Wu) = WP(u)W^*\},$$
$$\Gamma = \mathrm{Aut}(A) = \{W \in \Gamma(A); \ Wc = c\},$$

where A is a Jordan algebra (cf. Lemma 2).

A linear transformation D of A is called a *derivation of A* if

$$D(u \circ v) = Du \circ v + u \circ Dv$$

for all $u, v \in \mathcal{A}$. Substituting $u = v = c$ we get $Dc = 0$. It will be shown that $\mathcal{L}(Aut(\mathcal{A}))$ is equal to the vector space of all derivations of $\mathcal{A}$. Let $D \in \mathcal{L}(Aut(\mathcal{A}))$. Then there exists a curve $W = W(\tau) = \text{Id} + \tau D + O(\tau^2)$ in $Aut(\mathcal{A})$. Since $W = W(\tau)$ is an automorphism of $\mathcal{A}$, one has $W(u \circ v) = Wu \circ Wv$. We compare the coefficients of τ in order to see that D is a derivation. Conversely, let D be a derivation of $\mathcal{A}$. By induction we get

$$D^m(u \circ v) = \sum_{\nu=0}^{m} \binom{m}{\nu} D^\nu u \circ D^{m-\nu} v.$$

Therefore $W = W(\tau) = e^{\tau D}$ belongs to $Aut(\mathcal{A})$, hence $D \in \mathcal{L}(Aut(\mathcal{A}))$.

For linear transformations T_1 und T_2 we define the *Lie product* by

$$[T_1, T_2] := T_1 T_2 - T_2 T_1.$$

Lemma 7. *(a) $\mathcal{L}(\Gamma(\mathcal{A}))$ is the direct sum of the vector spaces $\mathcal{L}(Aut(\mathcal{A}))$ and $L(\mathcal{A}) = \{L(x); \ x \in \mathcal{A}\}$.*

(b) $\mathcal{L}(\Gamma(\mathcal{A}))$ and $\mathcal{L}(Aut(\mathcal{A}))$ are Lie algebras.

(c) $[L(\mathcal{A}), L(\mathcal{A})] \subset \mathcal{L}(Aut(\mathcal{A})), \quad [L(\mathcal{A}), \mathcal{L}(Aut(\mathcal{A}))] \subset L(\mathcal{A})$.

Proof. (a) Let $T = D + L(f)$, $D \in \mathcal{L}(Aut(\mathcal{A}))$. We consider the curve

$$W(\tau) = e^{\tau L(f)} e^{\tau D} \in \Gamma(\mathcal{A}).$$

In view of $W(0) = \text{Id}$, $\dot{W}(0) = L(f) + D = T$, we get $T \in \mathcal{L}(\Gamma(\mathcal{A}))$. Conversely, let $T \in \mathcal{L}(\Gamma(\mathcal{A}))$. There exists a curve $W = W(\tau) = \text{Id} + \tau T + O(\tau^2) \in \Gamma(\mathcal{A})$. Moreover, $\tau \mapsto W^*(\tau)$ is also twice continuously differentiable, with $W^*(\tau) = \text{Id} + \tau T^* + O(\tau^2)$. Differentiation of $P(Wu) = WP(u)W^*$ yields

(1) $$TP(u) + P(u)T^* = 2P(u, Tu).$$

Substituting $u = c$ shows $T^* = -T + 2L(f)$, with $f = Tc$. Hence $T^*c = f$. We apply (1) to c and get

(2) $$Tu^2 + P(u)f = 2P(u, Tu)c = 2Tu \circ u.$$

Denote the product in the mutation $\mathcal{A}_f$ by $\perp$. In (2) we substitute $u + v$ for u and compare the linear terms in u, v:

$$T(u \circ v) = Tu \circ v + u \circ Tv - u \perp v.$$

For $T = L(f) + D$ it follows that $D(u \circ v) = Du \circ v + u \circ Dv$, hence $D \in \mathcal{L}(Aut(\mathcal{A}))$.

The sum $\mathcal{L}(Aut(\mathcal{A})) + \{L(f); \ f \in \mathcal{A}\}$ is direct: $L(f) + D = 0$ immediately leads to $L(f)c + Dc = 0$, thus $f = 0$ and then $D = 0$.

(b) In view of (c) we only need to prove that $\mathcal{L}(Aut(\mathcal{A}))$ is a Lie algebra. $\mathcal{L}(Aut(\mathcal{A}))$ contains exactly the derivations of $\mathcal{A}$ and it is easy to see that $[D_1, D_2]$ is a derivation provided that D_1 and D_2 are derivations.

(c) By definition a linear map D belongs to $\mathcal{L}(Aut(\mathcal{A}))$ if and only if

$$(3) \qquad\qquad L(Dx) = DL(x) - L(x)D = [D, L(x)]$$

holds for every x. Hence $[L(\mathcal{A}), \mathcal{L}(Aut(\mathcal{A}))] \subset L(\mathcal{A})$ follows. For $D = [L(x), L(y)]$ we use Chapter III, §2, equation (3), and get

$$L(Du) = L(x \circ (y \circ u)) - L(y \circ (x \circ u)) = DL(u) - L(u)D.$$

Hence $[L(\mathcal{A}), \ L(\mathcal{A})] \subset \mathcal{L}(Aut(\mathcal{A}))$ follows.   □

Now assume that $\mathcal{A}$ is semisimple. Since $L(u)$ is self-adjoint with respect to the bilinear form $\tau(x, y)$, it follows from (3) that for any derivation D, $D + D^*$ commutes with every $L(u)$, $u \in \mathcal{A}$. To verify this, use

$$\tau(DL(u)x, y) - \tau(L(u)Dx, y) = \tau(L(Du)x, y),$$

and take adjoints in order to obtain

$$DL(u) - L(u)D = L(Du) = L(u)D^* - D^*L(u).$$

Now Lemma III.2 implies that $D + D^* = L(d)$, with d an element in the center $Z(\mathcal{A})$. Forming the trace in (3) we obtain $\tau(c, Du) = \mathrm{tr}L(Du) = 0$ and hence $D^*c = 0$. In view of $Dc = 0$ we get $L(d)c = d = 0$ and

$$D^* = -D$$

for every derivation of $\mathcal{A}$.

Without proof we quote the following theorem of N. Jacobson [Jac1].

Theorem 8. *Let $\mathcal{A}$ be a semisimple Jordan algebra. Then every derivation of $\mathcal{A}$ is a finite sum of transformations*

$$L(x)L(y) - L(y)L(x), \quad \text{with} \quad x, y \in \mathcal{A}.$$

The fundamental formula in Theorem 1 states that $P(u)$, $\det P(u) \neq 0$, belongs to $\Gamma(\mathcal{A})$ and we have seen in Lemma 6 that

$$V(u, v) = P((P(u)v^2)^{-1/2})P(u)\, P(v)$$

belongs to the group $Aut(\mathcal{A})$ of automorphisms of $\mathcal{A}$. Now we prove

Theorem 9. *Let A be a semisimple Jordan algebra.*

(a) The connected component of the identity of $\Gamma(A)$ is generated by $P(u)$, where u varies in a neighborhood of c.

(b) The connected component of the identity of $Aut(A)$ is generated by $V(u, v)$, where u, v vary in a neighborhood of c.

Proof. Since a connected topological group is generated by any neighborhood of the unit element, we only need to prove that $P(u)$ resp. $V(u, v)$ for u, v in a neighborhood of c generate a neighborhood of Id in $\Gamma(A)$ resp. in $Aut(A)$. Lemma 5 states that in a neighborhood of Id in $\Gamma(A)$ any element can be written in the form $P(u)V$, where V belongs to a neighborhood of the identity in $Aut(A)$. Therefore we only need to prove that the $V(u, v)$ generate a neighborhood of the identity in $Aut(A)$.

Let $\mathcal{L} = \mathcal{L}(Aut(A))$ be the Lie algebra of $Aut(A)$, i.e. the algebra of derivations of A. For a sufficiently small neighborhood U of zero in $\mathcal{L}$ the map

$$\varphi : U \to Aut(A), \quad \varphi(D) = e^D,$$

is a topological map onto a neighborhood of the identity. From Theorem 8 we get a basis

$$D_k = L(x_k)L(y_k) - L(y_k)L(x_k), \quad k = 1, 2, \ldots r,$$

of the vector space $\mathcal{L}$. For small U we define the real analytic map

$$\psi : U \to Aut(A), \quad \psi(D) = \prod_{k=1}^{r} V_k, \quad V_k = V(c + \lambda_k x_k, c + \lambda_k y_k),$$

if

$$(4) \qquad\qquad D = \sum_{k=1}^{r} \lambda_k D_k.$$

Let $|x|$ be a norm on the underlying vector space of A. For sufficiently small $\varepsilon > 0$ we consider the neighborhood U consisting of all D in the form (4), where $|\lambda_k| < 1$ and $|x_k| < \varepsilon, |y_k| < \varepsilon$. As in Lemma 6 it follows that

$$V_k = \text{Id} - 2\lambda_k^2 D_k + O(\varepsilon^3),$$

and hence

$$\psi(D) = \text{Id} - 2D + O(\varepsilon^3).$$

Considering $\varphi^{-1} \circ \psi : U \to U$ we obtain

$$(\varphi^{-1} \circ \psi)(D) = \varphi^{-1}(\text{Id} - 2D + O(\varepsilon^3)) = -2D + O(\varepsilon^3).$$

Hence the Jacobian of $\varphi^{-1} \circ \psi$ in zero does not vanish and consequently $\varphi^{-1} \circ \psi$ is a topological map onto U. Therefore ψ is surjective and the V_k generate a neighborhood of Id in $Aut(A)$. $\qquad\square$

§6. Direct sums

Let A be a semisimple Jordan algebra, with

$$A = A_1 \oplus A_2 \oplus \ldots \oplus A_r$$

the decomposition of A into simple subalgebras A_ν (see Theorem III.11). We know that the A_ν are the minimal ideals of A and that this decomposition is unique up to a permutation. Let $\Gamma_1(A)$ be the direct product of the structure groups, i.e. $\Gamma(A_1) \times \Gamma(A_2) \times \ldots \times \Gamma(A_r)$. Then the elements of $\Gamma_1(A)$ are the linear transformations $W_1 \times W_2 \times \ldots \times W_r$, with $W_\nu \in \Gamma(A_\nu)$, defined by

$$(W_1 \times W_2 \times \ldots \times W_r)x = W_1 x_1 + W_2 x_2 + \ldots + W_r x_r,$$

for $x = x_1 + \ldots + x_r$, $x_\nu \in A_\nu$. Now we consider the quadratic representations $P(x)$ resp. $P_\nu(x_\nu)$ and their determinants $\pi(x)$ resp. $\pi_\nu(x_\nu)$. By virtue of $A_\nu \circ A_\mu = \{0\}$ for $\nu \neq \mu$ we get

(1) $\quad P(x) = P_1(x_1) \times \ldots \times P_r(x_r), \quad$ where $\quad x = x_1 + \ldots + x_r,\ x_\nu \in A_\nu,$

and hence

(2) $$\pi(x) = \pi_1(x_1) \cdot \ldots \cdot \pi_r(x_r).$$

Theorem 5 now implies that $\Gamma_1(A)$ is a subgroup of $\Gamma(A)$. Another subgroup of $\Gamma(A)$ is defined as follows: For $\nu \neq \mu$ let $V_{\nu\mu}$ be the linear transformation defined as follows: $V_{\nu\mu}$ is the identity if A_ν and A_μ are not isomorphic. In the case that A_ν and A_μ are isomorphic and $\sigma : A_\nu \to A_\mu$ is a fixed isomorphism, then $V_{\nu\mu}$ maps $x = x_1 + \ldots + x_r$ to $y = y_1 + \ldots + y_r$, where $y_\nu = \sigma^{-1}(x_\mu)$, $y_\mu = \sigma(x_\nu)$ and $y_\varrho = x_\varrho$ for $\varrho \neq \nu$, $\varrho \neq \mu$. Let $\Gamma_0(A)$ be the subgroup of $\Gamma(A)$ generated by the $V_{\nu\mu}$ for $\nu \neq \mu$. If the isomorphisms between A_ν and A_μ are properly chosen then $\Gamma_0(A)$ will be a finite subgroup of $\Gamma(A)$.

Given $f = f_1 + \ldots + f_r \in A$ we consider the mutations A_f and A_{ν,f_ν}. Obviously we get

(3) $$A_f = A_{1,f_1} \oplus \ldots \oplus A_{r,f_r}$$

but it is not true in general that the A_{ν,f_ν} are minimal ideals in A_f.

Theorem 10. *Let A be a semisimple Jordan algebra. Then one has*

$$\Gamma(A) = \Gamma_0(A) \cdot \Gamma_1(A).$$

Proof. We start with an element $W \in \Gamma(\mathcal{A})$ and $f = W^* c$. Theorem 7 implies that $W : \mathcal{A}_f \to \mathcal{A}$ is an isomorphism of the mutation $\mathcal{A}_f$ onto $\mathcal{A}$. Let us consider the decompositions $\mathcal{A} = \mathcal{A}_1 \oplus \ldots \oplus \mathcal{A}_r$ of $\mathcal{A}$ and of $\mathcal{A}_f$ given by (3). Since $W : \mathcal{A}_f \to \mathcal{A}$ is an isomorphism, it follows that $W \mathcal{A}_{\nu, f_\nu}$ is an ideal in $\mathcal{A}$ and that

$$\mathcal{A} = W \mathcal{A}_{\nu, f_\nu} \oplus \ldots \oplus W \mathcal{A}_{\nu, f_\nu}.$$

The $W \mathcal{A}_{\nu, f_\nu}$ are minimal ideals in $\mathcal{A}$. Indeed, if one of these ideals is not minimal then we can decompose it into at least two minimal ideals. Hence the number of minimals ideals in $\mathcal{A}$ would be $> r$; a contradiction.

Since this decomposition of $\mathcal{A}$ is unique up to order, we find a permutation $\pi = \pi(\nu)$ of $\{1, \ldots, r\}$ such that

$$W \mathcal{A}_{\nu, f_\nu} = \mathcal{A}_{\pi(\nu)}.$$

W is an isomorphism of $\mathcal{A}_f$ onto $\mathcal{A}$. Hence the restriction of W to $\mathcal{A}_{\nu, f_\nu}$ is injective. There is an element V in $\Gamma_0(\mathcal{A})$ such that

$$VW \mathcal{A}_{\nu, f_\nu} = V \mathcal{A}_{\pi(\nu)} = \mathcal{A}_\nu \quad \text{for} \quad \nu = 1, \ldots, r.$$

If we denote the restriction of VW onto $\mathcal{A}_{\nu, f_\nu}$ by W_ν we get

$$VW = W_1 \times \ldots \times W_r$$

and each $W_\nu : \mathcal{A}_{\nu, f_\nu} \to \mathcal{A}_\nu$, is an isomorphism. Theorem 7 states that $W_\nu \in \Gamma(\mathcal{A}_\nu)$ and therefore VW belongs to $\Gamma(\mathcal{A}_1) \times \ldots \times \Gamma(\mathcal{A}_r) = \Gamma_1(\mathcal{A})$. $\square$

Notes

In the literature the so-called quadratic representation P does not play an important role, but in the present applications of Jordan algebras it stands in the center of investigations. The fundamental formula was conjectured first by N. Jacobson [8], who also proved it for the exceptional Jordan algebra [9]. In general, the fundamental formula is a consequence of a theorem by I.G. Macdonald [21], but the proof of this theorem is too complicated for a lecture. The present proof is based on a proof due to Ch. Hertneck [4] (in the case of formally real Jordan algebras), and follows a generalization due to E. Artin. Concerning the characterization of Jordan algebras by the quadratic representation see also [19].

The name "mutation" was proposed by H. Braun; Jacobson uses the name "isotope". The Jordan identity for mutations was conjectured by N. Jacobson [8]; it also follows from Macdonald's theorem.

The group $\Gamma(\mathcal{A})$ is related to some groups investigated by N. Jacobson [11]. The notion of similarity of Jordan algebras was proposed by E. Artin.

Editors' Notes

1. (i) The asterisk $*$ is used in various ways in this chapter. Sometimes it denotes a gradient (with respect to the trace form τ), sometimes the adjoint of a linear map, on the partner of an element of the structure group $\Gamma(\mathcal{A})$. For $\mathcal{A}$ semisimple, and $W \in \Gamma(\mathcal{A})$, both possible interpretations coincide due to Theorem 5. The alternative description of $\Gamma(\mathcal{A})$ in the semisimple case was moved from its original place to a point following Theorem 5 by the editors, since the result of this theorem is needed to obtain the description. The Lie group property of $\Gamma(\mathcal{A})$ was originally stated only for the case of a semisimple algebra.

(ii) The structure group has another important property:

If the identity $P(Wx) = WP(x)W^*$ holds, and if z is invertible, then $(Wz)^{-1} = W^{*-1}z^{-1}$.

The proof follows directly from $z^{-1} = P^{-1}(z)z$ and the definitions. This identity will be used later on; see Chapter VII.

(iii) The notions of an automorphism resp. a derivation make sense for arbitrary nonassociative algebras. For real or complex algebras, the automorphisms form a linear Lie group, with the derivations forming the associated Lie algebra. The proofs can be taken almost verbatim from the corresponding theorems in this chapter.

(iv) The proper setting for using differential operators over arbitrary base fields is based on the use of generic elements (and algebras that keep satisfying the defining identities over any base field extension). This was carried out in Braun and Koecher [BK].

(v) In §5 it was shown that every derivation of a semisimple Jordan algebra is skew-symmetric with respect to the trace form τ. One may add that every automorphism of a Jordan algebra leaves the trace form invariant. To see this, let W be an automorphism of $\mathcal{A}$. Then $\tau(Wx, Wy) = \mathrm{tr}L(W(x \circ y)) = \mathrm{tr}(WL(x \circ y)W^{-1}) = \tau(x, y)$ using $L(Wv)W = WL(v)$. This property will be used in Chapter VII.

(vi) Contrary to the statements in the author's notes, mutations are usually called *homotopes* by Jacobson and his school. The name *isotope* is reserved for mutations by invertible elements.

2. Koecher's remark about the quadratic representation not playing an important role in the literature (as of 1962) is not true any more, and the credit for this goes to Koecher and his school, and to McCrimmon. In fact, the two most important generalizations of the concept of a Jordan algebra, viz. Jordan triple systems (Meyberg [Mey]) and Jordan pairs (Loos [Lo2]) are based on axioms that generalize properties of the quadratic representation. The identity from Corollary 1(b) turns out to be just as important as the fundamental formula. The introduction of the structure group to the theory and applications of Jordan algebras is one of Koecher's most significant contributions.

Mutations of Jordan algebras occur naturally in the Koecher approach to ω–domains. Indeed, there is no distinguished base point, and a change of base points yields a mutation.

The fact that mutations of Jordan algebras are again Jordan is crucial for the Kantor-Koecher-Tits construction of Lie algebras from Jordan algebras; cf. Kantor [Ka], Koecher [Ko1], and Tits [Ti].

Chapter V. Complex Jordan Algebras

§1. Minimal polynomial and eigenvalues

Let $\mathcal{A}$ be a Jordan algebra with unit element c over an arbitrary field K of characteristic $\neq 2$. Let n be the finite dimension of $\mathcal{A}$ over K. Since the powers $c = u^0, u^1, u^2, \ldots, u^n$ of a given $u \in \mathcal{A}$ are linearly dependent, there is a nonzero polynomial $f(\tau)$ over K such that $f(u) = 0$. Consider the ideal $\mathcal{I}$ in $K[\tau]$ consisting of all polynomials $f(\tau)$ such that $f(u) = 0$. Since $\mathcal{I}$ is a principal ideal, there exists a uniquely determined normalized polynomial $p(\tau)$ such that $\mathcal{I} = (p(\tau))$. Hence, a given polynomial $f(\tau)$ satisfying $f(u) = 0$ is divisible by $p(\tau)$ and $p(\tau)$ has minimal degree among all the nonzero polynomials over K vanishing for u. Thus $p(\tau)$ is called the *minimal polynomial of* u. The degree of the minimal polynomial of u is equal to the dimension of the vector space $K[u]$ spanned by all the powers of u.

The minimal polynomial of the linear transformation $L(u)$ is the normalized polynomial $f(\tau)$ in $K[\tau]$ of minimal degree such that $f(L(u)) = 0$. In view of $L^k(u)c = u^k$ it follows that $0 = f(L(u))c = f(u)$. Therefore $p(\tau)$ is a divisor of $f(\tau)$.

A polynomial $q(\tau) \in K[\tau]$ is called an *eigenpolynomial* of u if

(EP.1) $q(\tau)$ is irreducible and normalized,

(EP.2) there exists $0 \neq v \in \mathcal{A}$ such that u and v commute and $q(u) \circ v = 0$.

Such a vector v is called an *eigenvector* of the eigenpolynomial $q(\tau)$. If an eigenpolynomial is linear, i.e. of the form $\tau - \zeta$ for some $\zeta \in K$, then ζ is called an *eigenvalue* of u. In this case we get $u \circ v = \zeta v$.

Lemma 1. *The eigenpolynomials of an element $u \in \mathcal{A}$ are exactly the irreducible normalized factors of the minimal polynomial of u. For every eigenpolynomial there exists an eigenvector in $K[u]$.*

Proof. Let f be a normalized irreducible factor of the minimal polynomial $p(\tau)$ of u, hence $p(\tau) = f(\tau)g(\tau)$. By virtue of $p(u) = 0$ it follows that

$0 = f(u) \circ g(u)$ and that $v := g(u)$ commutes with u. Therefore $f(\tau)$ is an eigenpolynomial and v is an eigenvector in $K[u]$.

On the other hand let $q(\tau)$ be an eigenpolynomial and let v be an eigenvector. Dividing $p(\tau)$ by $q(\tau)$ we get $p(\tau) = q(\tau)s(\tau) + r(\tau)$, and the degree of $r(\tau)$ is smaller than the degree of $q(\tau)$. It follows that

$$0 = p(u) \circ v = [q(u) \circ s(u)] \circ v + r(u) \circ v.$$

Since u and v commute, Chapter III, §5, equation (1), states that

$$[q(u) \circ s(u)] \circ v = s(u) \circ [q(u) \circ v] = 0.$$

Thus we get $r(u) \circ v = 0$. If $r(\tau) \neq 0$ then $q(\tau)$ and $r(\tau)$ are relatively prime, and $r(u) \circ v = q(u) \circ v = 0$ implies $v = 0$, a contradiction. Therefore $r(\tau) = 0$, and $q(\tau)$ is a divisor of $p(\tau)$. □

Corollary 1. *An element $u \in A$ is a divisor of zero in $K[u]$ if and only if $q(\tau) = \tau$ is an eigenpolynomial of u.*

Corollary 2. *An element $u \in A$ is invertible if and only if 0 is not a zero of the minimal polynomial of u.*

Proof. Because of Lemma III.5, we see that u^{-1} exists if and only if u is not a divisor of zero in $K[u]$. Hence we can apply Corollary 1. □

Now let us consider the eigenpolynomials of a nilpotent element $v \in A$. There is an $m > 0$ such that $v^m = 0$. If m is chosen minimal then $q(\tau) = \tau^m$ is the minimal polynomial, and vice versa. From Lemma 1 it therefore follows that v is nilpotent if and only if $g(\tau) = \tau$ is the only eigenpolynomial of v.

Lemma 2. *Let u and v be elements of an associative subalgebra of A and let v be nilpotent. Then u and $u + v$ possess the same eigenpolynomials.*

Proof. Let $p(\tau)$ be the minimal polynomial of u and let $v^m = 0$. Since u and v belong to a commutative and associative subalgebra, $p(u + v)$ is a linear combination of terms $u^i \circ v^j$. Now $p(u) = 0$ shows that $p(u + v) = v \circ w$ for some w, and $p^m(u + v) = 0$. Thus the minimal polynomial of $u + v$ is a divisor of $p^m(\tau)$. By virtue of Lemma 1 each eigenpolynomial of $u + v$ is an eigenpolynomial of u. In view of $u = (u + v) - v$ we can interchange the roles of u and $u + v$. □

§2. Minimal relations

A system $e_1, \ldots, e_r \in \mathcal{A}$ is called a *complete orthogonal system* if

$$e_i \circ e_j = \delta_{ij} e_i = \begin{cases} e_i & i = j, \\ 0 & i \neq j, \end{cases} \quad e_1 + \ldots + e_r = c, \quad e_i \neq 0.$$

In particular all the e_i are idempotents and $e_1, \ldots, e_r$ are linearly independent. Moreover, an application of Lemma III.1 shows that the e_i commute pairwise.

Given a non-empty subset $I \subset \{1, 2, \ldots, r\}$ we define

$$e_I := \sum_{i \in I} e_i$$

Thus e_I and the e_j, $j \notin I$, also form a complete orthogonal system. More generally any partition of $\{1, 2, \ldots, r\}$ gives rise to a complete orthogonal system.

We will say that for $u \in \mathcal{A}$ and a complete orthogonal system $e_1, \ldots, e_r$ an equation

$$(1) \qquad \sum_{k=1}^{r} p_k(u) \circ e_k = v$$

is a *relation of u with respect to* $e_1, \ldots, e_r$, if

(R.1) u and the e_i are contained in some associative subalgebra of $\mathcal{A}$,

(R.2) v is nilpotent,

(R.3) all the $p_k(\tau)$ are normalized irreducible polynomials in $K[\tau]$.

Obviously, the nilpotent element v then belongs to this associative subalgebra. It is called the *nilpotent part* of the relation. Collecting the e_i with the same $p_k(\tau)$ we get another relation with respect to a complete orthogonal system from (1), with the "coefficients" $p_k(\tau)$ pairwise different. In this case, (1) is called a *minimal relation of u*.

Multiplying (1) by e_i yields

$$(2) \qquad p_i(u) \circ e_i = v_i,$$

with $v_i = v \circ e_i$ nilpotent and belonging to the given associative subalgebra. It follows that $p_i^{m_i}(u) \circ e_i = 0$ holds for some $m_i > 0$. Let $f(\tau)$ be the product of the polynomials $p_i^{m_i}(\tau)$. Then one has $f(u) \circ e_i = 0$ for all i, and $f(u) = 0$. The minimal polynomial of u is therefore a divisor of $f(\tau)$ and Lemma 1 states that the eigenpolynomials of u occur among the p_i in relation (1).

Now we prove that any $u \in \mathcal{A}$ possesses a minimal relation.

Theorem 1. *Given $u \in A$ there exists a complete orthogonal system $c_1, \ldots, c_r$ in $K[u]$, and a minimal relation*

$$\sum_{k=1}^{r} p_k(u) \circ c_k = v.$$

The p_k are exactly the eigenpolynomials of u. Moreover, $v = 0$ holds if and only if $p_1(\tau) \cdot \ldots \cdot p_r(\tau)$ is the minimal polynomial of u.

Proof. Let

$$p(\tau) = \prod_{k=1}^{r} p_k^{\alpha_k}(\tau)$$

be the factorization of the minimal polynomial $p(\tau)$ of u, with normalized irreducible $p_k(\tau)$ in $K[\tau]$. The polynomials

$$\varphi_k(\tau) = \frac{p(\tau)}{p_k^{\alpha_k}(\tau)} = \prod_{l \neq k} p_l^{\alpha_l}(\tau)$$

and $p_k^{\alpha_k}(\tau)$ are relatively prime. Hence there exist polynomials $g_k(\tau)$ and $h_k(\tau)$ in $K[\tau]$ such that

$$1 = g_k(\tau)\varphi_k(\tau) + h_k(\tau)p_k^{\alpha_k}(\tau).$$

We put

$$j_k(\tau) = g_k(\tau)\varphi_k(\tau)$$

and see that each of the polynomials

$$j_k(\tau)(1 - j_k(\tau)), \quad j_k(\tau)j_l(\tau) \, (l \neq k), \quad 1 - \sum_{k=1}^{r} j_k(\tau)$$

is divisible by $p(\tau)$. This shows that

$$c_k := j_k(u), \quad k = 1, 2, \ldots, r,$$

is a complete orthogonal system in $K[u]$. Since $p(\tau)$ is a divisor of the polynomial $[p_k(\tau)j_k(\tau)]^{\alpha_k}$, each $p_k(u) \circ c_k$ is nilpotent. Since these elements belong to the associative and commutative subalgebra $K[u]$, their sum is nilpotent, and we get a minimal relation for u. At the same time it follows that $v = 0$ if every $\alpha_k = 1$. Conversely, let $v = 0$ in this minimal relation. Multiplying the equation by c_i we get $p_i(u) \circ c_i = 0$ and thus $p_1(u) \circ \ldots \circ p_r(u) \circ c_i = 0$ for all i, whence $p_1(\tau) \cdot \ldots \cdot p_r(\tau)$ is the minimal polynomial of u. $\square$

In the next section we will assume that the base field K is algebraically closed. Let us here briefly consider the case that K is the field $\mathbf{R}$ of real numbers. Given an element $u \in \mathcal{A}$ the minimal polynomial splits into irreducible factors $p_k(\tau)$ of degree one or two. Let $c_1, \ldots, c_r$ be a complete orthogonal system of u. Then we know that $p_k^{m_k}(u) \circ c_k$ is nilpotent for every $m_k > 0$. We choose m_k such that the degree of $p_k^{m_k}(\tau)$ is equal to 2 for each k, and we form

$$\sum_{k=1}^{r} p_k^{m_k}(u) \circ c_k.$$

Since this sum is nilpotent, we can rewrite it as a relation

$$u^2 - 2a \circ u + b = v,$$

with $v \in \mathbf{R}[u]$ nilpotent, and a and b linear combinations of $c_1, \ldots, c_r$ with coefficients in $\mathbf{R}$. Let S be the vector space over $\mathbf{R}$ spanned by the idempotents $c_1, \ldots, c_r$ and all the nilpotent elements in $\mathbf{R}[u]$. Then S is a subalgebra of $\mathbf{R}[u]$. The algebra $\mathbf{R}[u]$ is a quadratic extension of S in the sense that there exists $w \in \mathbf{R}[u]$, satisfying $w^2 \in S$, such that the elements of $\mathbf{R}[u]$ are exactly those of the form $a \circ w + b$, where $a, b \in S$.

§3. The minimal decomposition

For the rest of this chapter let K be the field $\mathbf{C}$ of complex numbers. Given $u \in \mathcal{A}$ the minimal polynomial $p(\tau)$ of u splits into linear factors $\tau - \xi_k$. It follows from Lemma 1 that the eigenpolynomials $q(\tau)$ of u are exactly the different ones among the polynomials $\tau - \xi_k$. We get $u \circ v = \xi v$ for some nonzero v which commutes with u if and only if ξ is a zero of the minimal polynomial of u. We call these zeros the *eigenvalues* of u.

Any minimal relation of u with respect to the orthogonal system $e_1, \ldots, e_r$ can be rewritten as a decomposition

$$(1) \qquad u = \sum_{k=1}^{r} \eta_k e_k + v, \quad v \quad \text{nilpotent},$$

where the η_k are the zeros of the linear polynomials p_k. Given a complete orthogonal system, we call (1) a *decomposition of u* if v is nilpotent and u and v belong to an associative subalgebra of $\mathcal{A}$. We prove that the different ones among the η_k are exactly the eigenvalues of u. Lemma 2 implies that we only need to demonstrate that the different η_k are the eigenvalues of

$$w = \sum_{k=1}^{r} \eta_k e_k.$$

Indeed, an arbitrary polynomial $f(\tau)$ satisfies $f(w) = \sum_{k=1}^{r} f(\eta_k)e_k$, and therefore $f(w) = 0$ if and only if $f(\eta_k) = 0$, $k = 1, \dots, r$.

If the η_k in (1) are mutually different then (1) is called a *minimal decomposition* of u. Collecting idempotents with equal coefficients we obtain a minimal decomposition from an arbitrary decomposition.

Theorem 1 now yields the existence of minimal decompositions:

Theorem 2. *Given $u \in A$ there exists a complete orthogonal system $c_1, \dots, c_r$ in $\mathbf{C}[u]$ such that*

$$u = \sum_{k=1}^{r} \xi_k c_k + v, \quad v \in \mathbf{C}[u] \quad nilpotent,$$

and the ξ_k are the eigenvalues of u. Moreover, $v = 0$ holds if and only if the minimal polynomial of u has only simple zeros.

The proof of Theorem 1 yields the

Remark. If all the eigenvalues of u are real numbers then the orthogonal system $c_1, \dots, c_r$ and the nilpotent part v can be chosen in $\mathbf{R}[u]$.

Theorem 3. *A minimal decomposition of $u \in A$ is uniquely determined up to the order of the complete orthogonal system.*

Proof. Let

$$u = \sum_k \eta_k e_k + v_1 \quad resp. \quad u = \sum_k \xi_k c_k + v_2$$

be an arbitrary minimal decomposition resp. the special decomposition from Theorems 1 and 2. We know that u and v_1 belong to an associative subalgebra and that $c_k = j_k(u) \in \mathbf{C}[u]$ and $v_2 \in \mathbf{C}[u]$. Moreover, the η_k are the eigenvalues of u. Therefore we assume without loss of generality that $\eta_k = \xi_k$. Using both decompositions we compute

$$j_l(u - v_1) = \sum_k j_l(\xi_k)e_k, \quad j_l(u - v_2) = \sum_k j_l(\xi_k)c_k.$$

For $i = 1, 2$, the elements u and v_i belong to an associative subalgebra, hence

$$j_l(u - v_1) = j_l(u) + w_1, \quad j_l(u - v_2) = j_l(u) + w_2, \quad w_2 \in \mathbf{C}[u],$$

with w_1 and w_2 nilpotent. It follows that

$$(2) \qquad c_l + w_1 = \sum_k j_l(\xi_k)e_k, \quad c_l + w_2 = \sum_k j_l(\xi_k)c_k.$$

The second equation shows that $j_l(\xi_k) - \delta_{lk}$ are the eigenvalues of the nilpotent element w_2. Since a nilpotent element has only 0 as an eigenvalue, we get

$$j_l(\xi_k) = \delta_{lk}.$$

(This formula also follows from the proof of Theorem 1.) Now the use of the first equation in (2) yields $c_l + w_1 = e_l$. Thus $c_l - e_l$ is nilpotent. Since w_1 and u belong to an associative subalgebra, c_l and e_l belong to the same subalgebra. Now $c_l = e_l$, and therefore uniqueness, is a consequence of the following proposition.   □

Proposition 1. *Let e_1 and e_2 be two idempotents in a commutative and associative algebra, and let $e_1 - e_2$ be nilpotent. Then $e_1 = e_2$.*

Proof. Let $(e_1 - e_2)^m = 0$ and m odd. Then we get

$$0 = e_1 + \sum_{k=1}^{m-1} \binom{m}{k} (-1)^k e_1 \circ e_2 - e_2$$

$$= e_1 - e_2 + e_1 \circ e_2 \cdot \sum_{k=0}^{m} \binom{m}{k} (-1)^k = e_1 - e_2.$$

□

§4. Applications of the minimal decomposition

Let $\mathcal{A}$ be a complex Jordan algebra with unit element c. As a first application of the minimal decomposition we prove

Lemma 3. *For every invertible $u \in \mathcal{A}$ there exists an element $u^{1/2} \in C[u]$ such that*

$$(u^{1/2})^2 = u, \quad C[u^{1/2}] = C[u].$$

Proof. Given a complex number λ we denote by $\lambda^{1/2}$ the value of $\sqrt{\lambda}$ such that $0 \leq \arg(\lambda^{1/2}) < \pi$. For invertible u we consider the minimal decomposition

$$u = \sum_k \xi_k c_k + v \quad \text{with } c_k, v \in C[u], \ v \text{ nilpotent}.$$

Corollary 2 shows that $\xi_k \neq 0$ for all k. Hence $\sum_k \xi_k c_k$ is invertible, and we obtain

$$u = \left(\sum_k \xi_k c_k\right) \circ (c + v_1),$$

with $v_1 \in \mathbf{C}[u]$ nilpotent. Now we define

$$u^{1/2} = \left(\sum_k \xi_k^{1/2} c_k\right) \circ \left(\sum_{m=0}^{\infty} \binom{1/2}{m} v_1^m\right) = \sum_k \xi_k^{1/2} c_k + v_2,$$

with $v_2 \in \mathbf{C}[u]$ nilpotent. Obviously we get $(u^{1/2})^2 = u$.   □

In general, $(u^2)^{1/2}$ is different from u. By virtue of $P(u) = P^2(u^{1/2})$ we see that the inverse of $u^{1/2}$ exists. Define

$$u^{-1/2} := (u^{1/2})^{-1}, \quad \text{if } \det P(u) \neq 0.$$

With a similar argument we can prove that the exponential function $u \mapsto e(u)$ maps the algebra $\mathcal{A}$ *onto* the set $\{x \in \mathcal{A}; \ \det P(x) \neq 0\}$. Indeed, let x be invertible, and

$$x = \left(\sum_k \xi_k c_k\right) \circ (c + v_1)$$

from the minimal decomposition. Choosing η_k such that $e^{\eta_k} = \xi_k$, and putting

$$u = \left(\sum_k \eta_k c_k\right) \circ \left(\sum_{m=1}^{\infty} \frac{(-1)^{m+1}}{m} v_1^m\right),$$

we get $e(u) = x$. Obviously, $u \mapsto e(u)$ is not a bijection for complex Jordan algebras. In order to discuss the equation $e(u) = e(\tilde{u})$ for $u, \tilde{u} \in \mathcal{A}$ we consider the minimal decomposition of u resp. $\tilde{u}$:

$$u = x + v, \quad x = \sum_k \xi_k c_k, \quad v \text{ nilpotent in } \mathbf{C}[u],$$

$$\tilde{u} = \tilde{x} + \tilde{v}, \quad \tilde{x} = \sum_l \tilde{\xi}_l \tilde{c}_l, \quad \tilde{v} \text{ nilpotent in } \mathbf{C}[\tilde{u}].$$

In view of $x, v \in \mathbf{C}[u]$ we obtain from Lemma IV.3,

(1) $e(u) = e(x) \circ e(v) = e(x) + v_1, \quad v_1 \text{ nilpotent in } \mathbf{C}[u],$

(2) $e(\tilde{u}) = e(\tilde{x}) \circ e(\tilde{v}) = e(\tilde{x}) + \tilde{v}_1, \quad \tilde{v}_1 \text{ nilpotent in } \mathbf{C}[\tilde{u}].$

Since v_1 and $e(u)$ resp. $\tilde{v}_1$ and $e(\tilde{u})$ belong to associative subalgebras, the equations (1) resp. (2) describe decompositions of $e(u) = e(\tilde{u})$. Passing to the minimal decomposition does not change the nilpotent part, and Theorem 3 states that $v_1 = \tilde{v}_1$, whence

(3) $$e(x) = e(\tilde{x}).$$

Now we apply Theorem IV.2 to (1) and (2) and we obtain

$$P(e(x))P(e(v)) = P(e(\tilde{x}))P(e(\tilde{v})).$$

In view of $\det P(e(x)) \neq 0$ it follows that $P(e(v)) = P(e(\tilde{v}))$. Lemma IV.4 now yields

$$e^{2L(v)} = e^{2L(\tilde{v})}.$$

Since v and $\tilde{v}$ are nilpotent, Theorem III.5 shows that $L(v)$ and $L(\tilde{v})$ are nilpotent. For a nilpotent linear transformation A one has

$$\sum_{m=1}^{\infty} \frac{(-1)^{m+1}}{m}(e^A - \mathrm{Id})^m = A,$$

and this implies $v = \tilde{v}$. To summarize:

$$e(u) = e(\tilde{u}) \quad \text{if and only if} \quad v = \tilde{v} \quad \text{and} \quad \Sigma_k e^{\xi_k} c_k = \Sigma_l e^{\tilde{\xi}_l} \tilde{c}_l.$$

The next result sharpens Lemma IV.5 in the complex case.

Lemma 4. *Any element $W \in \Gamma(\mathcal{A})$ can be written in the form*

$$W = P(u)V,$$

where V is an automorphism of $\mathcal{A}$ and $u = (Wc)^{1/2}$.

Proof. We substitute $x = c$ in $P(Wx) = WP(x)W^*$ in order to get $P(Wc) = WW^*$. Therefore we obtain $\det P(Wc) \neq 0$, and $(Wc)^{1/2}$ exists. Now it follows from $Vc = c$ that V is an automorphism of $\mathcal{A}$. □

There is another application of the minimal decomposition as follows.

Theorem 4. *Let $u, v \in \mathcal{A}$ and $\det(P(u)P(v)) \neq 0$. Then*

$$P(u)P(u^{-1} - v^{-1})P(v) = P(u - v)$$

and

$$(u^{-1} - v^{-1})^{-1} = u - P(u)(u - v)^{-1}$$

provided that $\det P(u - v) \neq 0$.

Proof. There exist $x, y \in \mathcal{A}$ such that

$$u = P(x)c = x^2 , \quad v = P(x)y.$$

Now the fundamental formula yields

$$P(u)P(u^{-1} - v^{-1})P(v) - P(u - v)$$
$$= P^2(x)P(P^{-1}(x)c - P^{-1}(x)y^{-1})P(P(x)y) - P(P(x)(c - y))$$
$$= \quad P(x)(P(c - y^{-1})P(y) - P(c - y))P(x),$$

and the first assertion follows from

$$P(c - y^{-1})P(y) = (\text{Id} - 2L(y^{-1}) + P(y^{-1}))P(y)$$
$$= P(y) - 2L(y^{-1})P(y) + \text{Id}$$
$$= P(y) - 2L(y) + \text{Id} = P(c - y)$$

(see Theorem III.12). In the same way the second statement can be shown to be equivalent to

$$(c - y^{-1})^{-1} = c - (c - y)^{-1}.$$

But this is true, because $\mathbf{C}[y]$ is associative.   □

§5. The eigenvalues of $L(u)$ and $P(u)$

Let $u \in \mathcal{A}$ and let

$$u = \sum_k \xi_k c_k + v , \; v \in \mathbf{C}[u] \quad \text{nilpotent},$$

be the minimal decomposition of u with respect to the complete orthogonal system c_k in $\mathbf{C}[u]$ (see Theorem 2). We will consider the ring $\Lambda = \Lambda_u$ generated by the transformations $L(w), w \in \mathbf{C}[u]$. Since the elements of $\mathbf{C}[u]$ commute pairwise, Λ is a commutative ring of linear transformations. Given $w \in \mathbf{C}[u]$ we get

$$w = \sum_k \eta_k c_k + v' , \quad v' \quad \text{nilpotent},$$

and hence

$$L(w) = \sum_k \eta_k L(c_k) + L(v').$$

By virtue of $c_k, v' \in \mathbf{C}[u]$, the transformations $L(c_k)$ and $L(v')$ belong to Λ. Therefore Λ is generated by the $L(c_k)$ together with the transformations

$L(v')$, with $v' \in \mathbf{C}[u]$ nilpotent. Let N be the radical of Λ and let Λ' denote the subring of Λ generated by the transformations $L(c_k)$, then it follows immediately that

$$\Lambda = \Lambda' \oplus N$$

is a direct sum of the underlying vector spaces.

In order to derive results about the structure of the ring Λ' we define the transformations

$$C_k = P(c_k) \neq 0, \; C_{kl} = 4L(c_k)L(c_l) = C_{lk} \quad \text{for} \quad k \neq l,$$

which are obviously contained in Λ'.

Lemma 5. *The transformations C_k and those $C_{kl}(k < l)$ which are not zero form a complete orthogonal system in Λ'.*

Proof. We recall equation (1) in Chapter III, §1:

(1) $$2L(x \circ y)L(x) + L(x^2)L(y) = 2L(x)L(y)L(x) + L(x^2 \circ y)$$

for $x, y \in \mathcal{A}$. Substituting $x = c_k$, $y = c_l$ we obtain

(2) $$P(c_k)L(c_l) = \delta_{kl} P(c_k).$$

If $k \neq l$ we get $2L^2(c_k)L(c_l) = L(c_k)L(c_l)$. We multiply this formula by $8L(c_l)$ and obtain $C_{kl}^2 = C_{kl}$ for $k \neq l$. Theorem IV.2 immediately leads to $P(c_k \circ c_l) = P(c_k)P(c_l)$. This shows that the transformations C_k and $C_{kl}(k \neq l)$ are idempotent (or zero), and that the C_k are orthogonal.

Now we substitute $x = c_k + c_l$, $y = c_j$ in (1) for mutually distinct k, l, j. Thus we obtain

$$0 = P(c_k + c_l)L(c_j) = [P(c_k) + 4L(c_k)L(c_l) + P(c_l)]L(c_j).$$

Hence (2) shows that

(3) $$L(c_k)L(c_l)L(c_j) = 0$$

holds. Therefore we get $C_{kl}C_{ji} = 0$ if $(k, l) \neq (j, i)$. Moreover (2) leads to

$$C_k C_{lj} = 4P(c_k)L(c_l)L(c_j) = 4\delta_{kl}\delta_{kj}P(c_k) = 0$$

if $l < j$. We have proved that the transformations C_k and $C_{kl} \neq 0, k < l$, form an orthogonal system. In order to demonstrate completeness we use $c = \sum_k c_k$ and get

$$L(c_k) = L(c_k)L(c) = \sum_l L(c_k)L(c_l) = L^2(c_k) + \frac{1}{4}\sum_{l\neq k} C_{kl},$$

hence

(4)
$$L(c_k) = C_k + \frac{1}{2}\sum_{l\neq k} C_{kl}.$$

Summation over k yields

$$Id = \sum_k C_k + \sum_{l<k} C_{kl}.$$

$\square$

Now we decompose $L(u)$ and $P(u)$ with respect to the orthogonal system of Λ'. We multiply (4) by the eigenvalue ξ_k of u. Then summation over k leads to

(5)
$$L(u) = \sum_k \xi_k C_k + \sum_{k<l} \frac{1}{2}(\xi_k + \xi_l)C_{kl} + L(v).$$

Moreover we obtain

$$u^2 = \sum_k \xi_k^2 c_k + v_2 , \quad v_2 \in \mathbf{C}[u] \quad \text{nilpotent.}$$

A combination of this formula and (5) yields

$$L(u^2) = \sum_k \xi_k^2 C_k + \sum_{k<l} \frac{1}{2}(\xi_k^2 + \xi_l^2)C_{kl} + L(v_2),$$

$$L^2(u) = \sum_k \xi_k^2 C_k + \sum_{k<l} \frac{1}{4}(\xi_k + \xi_l)^2 C_{kl} + V,$$

where $V \in \Lambda$ is nilpotent. Finally it follows that

(6)
$$P(u) = \sum_k \xi_k^2 C_k + \sum_{k<l} \xi_k\xi_l C_{kl} + W,$$

where $W \in \Lambda$ is nilpotent.

Since Λ is in particular a Jordan algebra, (5) and (6) give a decomposition with respect to a complete orthogonal system; now Theorems 2 and 3 yield

Theorem 5. *Let $u \in A$ and let the ξ_k be the mutually distinct eigenvalues of u. Then the eigenvalues of $L(u)$ resp. $P(u)$ are the distinct ones among the numbers*

$$\xi_k \quad \text{and} \quad \frac{1}{2}(\xi_k + \xi_l), \quad \text{if} \quad k < l \text{ and } C_{kl} \neq 0,$$

resp.

$$\xi_k^2 \quad \text{and} \quad \xi_k\xi_l, \quad \text{if} \quad k < l \text{ and } C_{kl} \neq 0.$$

Obviously, the eigenvalues of a linear transformation T, considered as an element of the Jordan algebra $\mathcal{A}$, are the same as the usual eigenvalues of the linear transformation T, because both of them are the zeros of the minimal polynomial of T.

Since the determinant of a linear transformation is the product of certain powers of its eigenvalues, we get

$$\det P(u) = \prod_k \xi_k^{\alpha_k} , \quad \text{if} \quad u = \sum_k \xi_k c_k + v,$$

with positive integers α_k. Since $u - \lambda c$ has the eigenvalues $\xi_k - \lambda$, it follows that the eigenvalues of $u \in \mathcal{A}$ are the distinct zeros of the polynomial $q(\tau) = \det P(u - \tau c)$.

§6. The embedding of real Jordan algebras

Let $\mathcal{A} = (X, \circ)$ be a real Jordan algebra. The extension of the base field yields a vector space $X(i)$ over the field $\mathbf{C}$ of complex numbers:

$$X(i) = \{x + iy \; ; \; x, y \in X\},$$

where the vector space operations are defined in the usual way. Any linear transformation T of X can be extended to a linear transformation of $X(i)$ via $Tz = Tx + iTy$ for $z = x + iy$. Now we define a composition in $X(i)$ by

$$(x + iy) \circ (u + iv) = [x \circ u - y \circ v] + i[(y \circ u) + (x \circ v)].$$

If we define

$$L(z) = L(x) + iL(y) \quad \text{for} \quad z = x + iy \in X(i),$$

we get $z \circ w = L(z)w$ for $z, w \in X(i)$. Any basis of X over $\mathbf{R}$ is also a basis of $X(i)$ over $\mathbf{C}$. The linear transformation $L(x)$, $x \in X$, is then represented by a matrix for which the elements are linear in the components of x with respect to this basis. We get the representation of $L(z)$ if we substitute the components of z for those of x. Hence $L(z)$ is linear in z and therefore

$$L(z)w = L(w)z , \quad L(z)L(z^2) = L(z^2)L(z)$$

hold for $z, w \in X(i)$. Thus we have constructed a Jordan algebra $\mathcal{A}(i) = (X(i), \circ)$. The given real Jordan algebra $\mathcal{A}$ is embedded into $\mathcal{A}(i)$. If $\mathcal{A}$ contains a unit element c then c turns out to be the unit element of $\mathcal{A}(i)$, too.

The algebra $\mathcal{A}(i)$ admits an involution $z = x + iy \mapsto \overline{z} = x - iy$ such that for example $\overline{z \circ w} = \overline{z} \circ \overline{w}$. But in general z and $\overline{z}$ do not commute.

It is easy to see that $z = x + iy$ and $\bar{z}$ commute if and only if x and y commute. Moreover, $z = x + iy$ belongs to the center of $\mathcal{A}(i)$ if and only if x and y belong to the center of $\mathcal{A}$. Thus, if $\mathcal{Z}$ denotes the center of $\mathcal{A}$ then $\mathcal{Z}(i)$ is the center of $\mathcal{A}(i)$.

We have bilinear forms (see Chapter III, §2)

$$\tau(x, y) = \mathrm{tr} L(x \circ y), \quad \tilde{\tau}(z, w) = \mathrm{tr} L(z \circ w),$$

on $\mathcal{A}$ resp. $\mathcal{A}(i)$. If $= x + iy$, $w = u + iv$ we have

(1) $\qquad \tilde{\tau}(z, w) = \tau(x, u) - \tau(y, v) + i(\tau(x, v) + \tau(y, u))$.

Concerning the radicals of $\mathcal{A}$ and $\mathcal{A}(i)$ we therefore get

Theorem 6. $rad \mathcal{A}(i) = (rad \mathcal{A})(i)$.

Corollary 3. $\mathcal{A}(i)$ *is semisimple if and only if $\mathcal{A}$ is semisimple.*

Later on we will need the following

Lemma 6. *Let $\mathcal{A}$ be a real Jordan algebra with unit element c and let $a \in \mathcal{A}$, with $\det P(a) \neq 0$. Then in any neighborhood of c there exists an element w such that $\det L(P(w)a) \neq 0$.*

Proof. Since $\det L(P(u)a)$ is a polynomial in u, it is sufficient to prove that this polynomial is not identically zero. This, in turn, is proved by showing that $\det L(P(u)a)$ is not identically zero in the complex extension $\mathcal{A}(i)$ of $\mathcal{A}$. But this follows from $P(a^{-1/2})a = \mathrm{Id}$. $\qquad \square$

The complex minimal decomposition for $u \in \mathcal{A}$ in $\mathcal{A}(i)$ gives rise to a decomposition of u in $\mathcal{A}$. Since the minimal polynomial of u is real, the eigenvalues of u are real or complex conjugate by pairs. Let λ_k be the real and $\xi_l, \bar{\xi}_l$ be the non-real distinct eigenvalues of u. Then we get a minimal decomposition in $\mathcal{A}(i)$:

$$u = \sum_k \lambda_k c_k + \sum_l (\xi_l e_l + \bar{\xi}_l e_l') + v,$$

where the complete orthogonal system c_k, e_l, e_l' and the nilpotent part v belong to $\mathbf{C}[u]$. Since the eigenvalues of $\bar{u}$ are the same as those of u, it follows from Theorem 3 that v and the c_k are real, and $e_l' = \bar{e}_l$. Hence $u \in \mathcal{A}$ possesses a real decomposition

(2) $\qquad u = \sum_k \lambda_k c_k + \sum_l (\xi_l e_l + \bar{\xi}_l \bar{e}_l) + v$.

Here the numbers $\lambda_k = \bar{\lambda}_k$, ξ_l and $\bar{\xi}_l$ are the eigenvalues of u, while c_k, e_l and $\bar{e}_l$ form a complete orthogonal system, and $c_k, v, \xi_l e_l + \bar{\xi}_l \bar{e}_l \in \mathbf{R}[u]$.

Equation (2) has some applications; for example

Lemma 7. *Given a real Jordan algebra A one has*

$$\{x^2 \; ; \; x \in A, \; \det P(x) \neq 0\} = \{e(y) \; ; \; y \in A\}.$$

Proof. Let Q resp. E be the set on the left resp. right hand side of the statement. From $e(y) \circ e(y) = e(2y)$ it follows that $E \subset Q$. Given $u = x^2 \in Q$ we apply the decomposition (2) to x. Thus we get

$$u = \left(\sum_k \lambda_k^2 c_k + \sum_l (\xi_l^2 e_l + \bar{\xi}_l^2 \bar{e}_l) \right) \circ (c + v),$$

where $c_k, e_l, \bar{e}_l, v$ belong to some associative subalgebra, and v is nilpotent. For a solution η_l of $e^{\eta_l} = \xi_l$ consider

$$y = 2 \left(\sum_k \log |\lambda_k| c_k + \sum_l (\eta_l e_l + \bar{\eta}_l \bar{e}_l) \right) \circ \sum_{m=1}^{\infty} \frac{(-1)^{m+1}}{m} v^m.$$

Obviously, y is real, and it satisfies $e(y) = x^2 = u$. Hence $Q \subset E$ follows. $\square$

Lemma 8. *Let A be a real Jordan algebra and let C be a connected component of the set $\{u \in A \; ; \; \det P(u) \neq 0\}$. Then there is an element $d \in C$ with $d^2 = c$.*

Proof. Given $u \in C$ we have the decomposition (2) with all $\lambda_k \neq 0$, $\xi_l \neq 0$. It is obvious that there is a curve in C joining u and

$$d = \sum_k \varepsilon_k c_k + \sum_l (e_l + \bar{e}_l),$$

where $\varepsilon_k = +1$ or -1. Hence $d^2 = c$ follows. $\square$

Notes

The minimal decomposition was given by P. Jordan, J. von Neumann and E. Wigner [15] in the case of formally real Jordan algebras. The exposition here follows E. Artin; (cf. [0]).

Editors' Notes

1. (i) Many of the arguments in §1, rely on the power-associativity of $\mathcal{A}$, which guarantees that the map $K[\tau] \rightarrow K[u]$, $f(\tau) \mapsto f(u)$, is always a homomorphism.

(ii) Theorem 1 is essentially the consequence of a theorem on the structure of commutative and associative algebras, applied to $K[u]$. See Braun and Koecher [BK], Kap. 1, §§3 and 4.

(iii) The projections C_k and C_{kl} of Lemma 5 yield the Peirce decomposition with respect to the complete orthogonal system $c_1, \ldots, c_r$; see [BK], Kap. 8.

(iv) To verify that the complexification of a real Jordan algebra is again Jordan, one may argue as follows: By polarization (and since char $\mathbf{R} \neq 2, 3$), the Jordan identity is equivalent to the multilinear identity

$$(uv)(xy) + (xv)(uy) + (xu)(vy) = v((xu)y) + u((xv)y) + x((uv)y).$$

This identity (and commutativity) is clearly preserved when the base field is extended. The author's argument (which is not carried out in detail) is based on a similar observation.

2. For a different approach to the (generic) minimal polynomial see Braun and Koecher [BK], and also Faraut and Korányi [FKo], Chap. II. This monograph also contains a quick proof of the existence of a minimal relation ("spectral decomposition"), and its special properties, in the case of formally real Jordan algebras.

Chapter VI. Jordan Algebras and Omega Domains

Throughout this chapter the ground field is the field $\mathbf{R}$ of real numbers. We have seen in Chapter II that we can associate a Jordan algebra to a given ω-domain. In this chapter we will construct an ω-domain for a given semisimple Jordan algebra, and investigate the correspondence.

§1. The ω-domain of a Jordan algebra

Let $\mathcal{A} = (X, \circ)$ be a semisimple Jordan algebra over $\mathbf{R}$ and let $P(x)$ be its quadratic representation. Putting

$$c = c_{\mathcal{A}} = \text{ unit element of } \mathcal{A},$$
$$\omega(y) = \omega_{\mathcal{A}}(y) = \sqrt{|\det P(y)|},$$
$$X_{\mathcal{A}} = \{x \in \mathcal{A}; \det P(x) \neq 0\},$$
$$Y = Y_{\mathcal{A}} = \text{ the connected component of } c_{\mathcal{A}} \text{ in } X_{\mathcal{A}},$$

we get a triple (Y, ω, c). We will prove that this triple is an ω-domain (see Chapter II, §1). Since $P(y)$ is homogeneous of degree two, the pair (Y, ω) satisfies the axioms (D.1) and (D.2a)–(D.2c) of an ω-domain. In order to check axiom (D.2d) we have to consider the bilinear form $\Delta_y^u \Delta_y^v \log \omega(y)$ for $y \in Y$. Lemma IV.1 states that

$$(1) \qquad \Delta_y^u \Delta_y^v \log \omega(y) = -\tau(P^{-1}(y)u, v),$$

where $\tau(u, v) = \operatorname{tr} L(u \circ v)$ is the bilinear form associated with the Jordan algebra $\mathcal{A}$. Since $\mathcal{A}$ is semisimple and $\det P(y) \neq 0$ for $y \in Y$, the bilinear form (1) is non-singular. This shows that (D.2d) holds, and furthermore that the bilinear form $\sigma(u, v)$ associated with the triple (Y, ω, c) is equal to $\tau(u, v)$. It follows from (1) that

$$H(y) = P^{-1}(y) \quad \text{for} \quad y \in Y.$$

Moreover Lemma IV.1 yields

$$y^\sharp = y^{-1} \quad \text{for} \quad y \in Y.$$

In view of $P(y^{-1}) = P^{-1}(y)$, the point y^{-1} belongs to Y for every $y \in Y$. Hence the map $Y \to Y$, $y \mapsto y^{-1}$, is a bijection.

In order to prove axiom (D.3) we have to consider the group $\Sigma(Y, \omega)$ of all non-singular linear transformations W of X fulfilling $WY = Y$ and $\omega(Wy) = |\det W| \cdot \omega(y)$ for $y \in Y$. Since Y is open, the second condition is equivalent to $\det P(Wx) = \gamma \cdot \det P(x)$ for $x \in X$ by Theorem IV.5. The same theorem yields

$$\Sigma(Y, \omega) = \{W \in \Gamma(\mathcal{A}); \ WY = Y\}.$$

Now we can show that $H(u) = P^{-1}(u)$ belongs to $\Sigma(Y, \omega)$ for $u \in Y$. Consider the open set $Y_1 = H(u)Y$. We get $Y_1 \cap Y \neq \emptyset$; indeed, we have $u^{-1} \in Y$ and $u^{-1} = H(u)u \in Y_1$. On the other hand we obtain $Y_1 \cap \partial Y = \emptyset$, as follows from the fundamental formula for $\mathcal{A}$: Let $x = H(u)y = P^{-1}(u)y \in Y_1$. Then $\det P(x) = (\det P(u))^{-2} \det P(y) \neq 0$ and hence $Y_1 = H(u)Y \subset Y$ follow. Replacing u by u^{-1} we get $H(u)Y = Y$, and therefore $H(u) \in \Sigma(Y, \omega)$. Summarizing we obtain

Theorem 1. *Let $\mathcal{A}$ be a semisimple Jordan algebra over $\mathbf{R}$. Then the triple $(Y_\mathcal{A}, \omega_\mathcal{A}, c_\mathcal{A})$ is an ω–domain on the underlying vector space of $\mathcal{A}$. In particular*

$$H(y) = P^{-1}(y), \quad y^\sharp = y^{-1}, \quad \Sigma(Y_\mathcal{A}, \omega_\mathcal{A}) = \{W \in \Gamma(\mathcal{A}); \ WY = Y\}.$$

The bilinear forms associated with $\mathcal{A}$ and with the ω–domain are equal.

As an abbreviation, the ω–domain $(Y_\mathcal{A}, \omega_\mathcal{A}, c_\mathcal{A})$ is called *the ω–domain of the Jordan algebra $\mathcal{A}$.*

Theorem 2. *There exists a connected neighborhood V of c in $Y_\mathcal{A}$ such that the group generated by the $P(u)$, $u \in V$, acts transitively on every connected component of $X_\mathcal{A}$.*

Proof. Let U be a sufficiently small connected neighborhood of the unit element c such that every $u \in U$ can be written in the form $u = v^2$, with v belonging to a small connected neighborhood V of c (see Chapter IV, §4). Denote by Π the subgroup of $\Sigma = \Sigma(Y_\mathcal{A}, \omega_\mathcal{A})$ generated by the $P(v), v \in V$. Then Π is a connected topological group (with the topology induced from Σ). Given $a \in X_\mathcal{A}$ we consider the orbit

$$\Pi a = \{Wa; \ W \in \Pi\}$$

of the point a. As is well-known, $\Pi a = \Pi b$ if and only if $\Pi a \cap \Pi b \neq \emptyset$ for given $a, b \in Y$. Denote by C a connected component of $X_\mathcal{A}$.

Proposition 1. *For every $a \in C$ one has $\Pi a \subset C$.*

Proof. Since Π is connected, Πa is also connected. Thus $a \in C \cap \Pi a$ implies $\Pi a \subset C$. □

Proposition 2. *For every $a \in C$ the orbit Πa is open in C.*

Proof. It follows from Lemma IV.5 that the determinant $\det L(P(v)a)$ is not zero for some $v \in V$. Therefore we find a point $b \in \Pi a$ such that $\det L(b) \neq 0$. Now consider the map $v \mapsto P(v)b$ for $v \in V$. At the point $v = c$ the Jacobian of this map is $\det 2L(b) \neq 0$. Therefore $v \mapsto P(v)b$ maps a neighborhood of c onto a neighborhood of b. Since the image is contained in $\Pi b = \Pi a$, we get an open, non-empty subset A of Πa. For arbitrary $x \in \Pi a$, $x = Wb$, with $W \in \Pi$ and $b \in A$, the set WA is a neighborhood of x, and $WA \subset W\Pi a = \Pi a$. □

Propositions 1 and 2 show that C is the union of the distinct (open) orbits. Since C is connected, we get $C = \Pi d$ for every $d \in C$. Hence Theorem 2 has been proved. □

Corollary 1. *The group $\Sigma(Y_A, \omega_A)$ of the ω–domain Y_A associated with the semisimple Jordan algebra A acts transitively on Y_A.*

Corollary 2. *Given $u \in A$ with $\det P(u) \neq 0$ there exist $W \in \Sigma(Y_A, \omega_A)$ and an element d in the connected component of u such that $u = Wd$ and $d^2 = c$.*

Proof. According to Lemma V.7 there is an element d in the component of u which satisfies $d^2 = c$. □

Corollary 3. *The map $x \mapsto x^{-1}$ is a bijection of each connected component of X_A.*

Proof. Indeed, $x \mapsto x^{-1}$ maps a component into a connected subset of X_A, and Corollary 2 shows that the component contains the fixed point d of this mapping. □

Theorem 3. *Let C be a connected component of X_A and $f \in A$. Then*

(a) $f \in C$ if and only if $C = Y_B$, with B the semisimple mutation A_f of A.

(b) If $Y_B = Y_C$ for semisimple mutations B and C of A then B and C are isomorphic.

Corollary 4. *Every component of X_A is an ω-domain.*

Proof. (a) Let $f \in C$. Then Corollary 3 shows $f^{-1} \in C$. Now we consider the mutation $B = A_f$ of A, with unit element f^{-1} and quadratic representation $P_f(u) = P(u)P(f)$ (see Chapter IV, §§2 and 3). The connected component Y_B of the unit element of B is equal to the connected component C of X_A because of $X_A = X_B$.

Vice versa, let $C = Y_B$, where $B = A_f$. Since the unit element f^{-1} of B belongs to $Y_B = C$, we get $f \in C$ from Corollary 3.

(b) Let $C' = Y_B = Y_C$, where $B = A_f$ and $C = A_g$. Since B and C are semisimple, we obtain $\det(P(f)P(g)) \neq 0$ and therefore $X_B = X_C = X_A$. In view of $f^{-1} \in C', g^{-1} \in C'$, and since C' is a component of X_A, Theorem 2 and Corollary 3 yield that there is $W \in \Gamma(A)$ satisfying $f = Wg$. Now Theorem IV.7 shows that B and C are isomorphic.   □

Sometimes we are interested in the images of A resp. X_A under the maps $x \mapsto e(x)$ resp. $x \mapsto x^2$. Here we have

Theorem 4. *Given a semisimple Jordan algebra A one has*

$$\{x^2;\ x \in X_A\} = \{e(x);\ x \in A\} \subset Y_A.$$

Proof. In view of Lemma V.7 we only need to prove that $E := \{e(x);\ x \in A\}$ is a subset of Y_A. Clearly E is connected and $c = e(0) \in E$. Lemma IV.4 states that $\det P(e(x)) \neq 0$, and therefore E does not contain a point of the boundary of Y_A. Hence $E \subset Y_A$ follows.   □

We saw in Theorem 2 that a certain subgroup of $\Sigma(Y_A, \omega_A)$, which is generated by some transformations $P(u)$, acts transitively on Y_A. We now consider the group $\Pi(A)$ generated by the linear transformations $P(u)$, with $u \in A$ and $\det P(u) \neq 0$. Moreover $Aut(A)$ denotes the group of automorphisms of the Jordan algebra A.

Theorem 5. *The group $\Pi(A)$ is a subgroup of $\Sigma(Y_A, \omega_A)$ and we get*

$$\Sigma(Y_A, \omega_A) = Aut(A) \cdot \Pi(A) = \Pi(A) \cdot Aut(A).$$

Proof. In order to show that $P(u) \in \Sigma(Y_A, \omega_A)$, we only need to prove $P(u)Y_A = Y_A$, according to Theorem 1. In view of $\det P(u) \neq 0$, the set $P(u)Y_A$ is open and connected. The fundamental formula implies that $P(u)Y_A$ does not contain a point of the boundary of Y_A. The point $P(u)c = u^2$ belongs to $P(u)Y_A$ and to Y_A (see Theorem 4). Hence $P(u)Y_A \subset Y_A$, and analogously $P^{-1}(u)Y_A \subset Y_A$ follows.

In the same way one demonstrates that $Aut(\mathcal{A})$ is a subgroup of $\Sigma(Y_A, \omega_A)$.

Given $W \in \Sigma(Y_A, \omega_A)$, consider Wc. Since $\Pi(\mathcal{A})$ acts transitively on Y_A, there is $W_1 \in \Pi(\mathcal{A})$ such that $W_1 Wc = c$. Hence $W_1 W$ is an automorphism of $\mathcal{A}$, according to Lemma IV.2. Similarly, there is $W_2 \in \Pi(\mathcal{A})$ such that $W_2 W^{-1} c = c$, hence $W W_2^{-1} \in Aut(\mathcal{A})$. □

For a semisimple Jordan algebra $\mathcal{A}$ we have an ω–domain (Y_A, ω_A, c_A); we denote this triple by $\Omega(\mathcal{A})$:

$$\Omega(\mathcal{A}) = (Y_A, \omega_A, c_A).$$

The mapping $\mathcal{A} \to \Omega(\mathcal{A})$ maps the set of semisimple Jordan algebras on a given vector space X to the set of ω–domains on X.

§2. The Jordan algebra of an ω–domain

In Theorem II.4 we described the construction of a Jordan algebra from a given ω–domain (Y, ω, c) on a vector space X. Let us recall this: Starting with the linear transformation $H(y)$ defined by

$$\Delta_y^u \Delta_y^v \log \omega(y) = -\sigma(H(y)u, v), \quad \text{resp.} \quad H(y) = -\frac{\partial y^\sharp}{\partial y},$$

we obtain a linear transformation $L(x)$ from the linear term in the power series expansion of $H(y)$ around $y = c$:

$$H(c + x) = \text{Id} - 2L(x) + \dots .$$

Theorem II.4 shows that $L(x)$ defines a Jordan algebra with unit element c on the vector space X such that the exponential function $e(x)$ of this Jordan algebra satisfies

$$e^\sharp(x) = e(-x).$$

From Lemma IV.3 we know that the exponential function also satisfies

$$e(-x) = e^{-1}(x),$$

and that $x \mapsto e(x)$ maps a neighborhood of 0 onto a neighborhood of the unit element c. Hence

$$y^\sharp = y^{-1}$$

holds in a neighborhood of c in Y. The definition of the inverse in a Jordan algebra yields

$$y \circ y^{\sharp} = c, \quad L(y)L(y^{\sharp}) = L(y^{\sharp})L(y)$$

in this neighborhood. Since $y^{\sharp}$ is real-analytic in $y \in Y$, these identities hold for all points of Y, and it follows with Theorem III.12 that

(1) $\qquad y \in Y$ implies $\det P(y) \neq 0$ and $y^{\sharp} = y^{-1}$.

By virtue of $H(y) = -\frac{\partial y^{\sharp}}{\partial y}$ and $\frac{\partial y^{-1}}{\partial y} = -P^{-1}(y)$ (see Chapter IV, §1) we find

(2) $\qquad\qquad H(y) = P^{-1}(y)$ for all $y \in Y$.

Thus equation (6) in Chapter II, §1, yields

(3) $\qquad\qquad \omega(y) = \sqrt{|\det P(y)|}$.

Now let $\sigma(x,y)$ resp. $\tau(x,y)$ denote the bilinear form of the ω–domain resp. of the Jordan algebra $\mathcal{A}$. It follows from (2), (3), the definition of $H(y)$ and Lemma IV.1 that

$$\Delta_y^u \Delta_y^v \log \omega(y) = -\sigma(H(y)u, v) = -\sigma(P^{-1}(y)u, v),$$
$$= -\frac{1}{2}\Delta_y^u \Delta_y^v \log \det P(y) = -\tau(P^{-1}(y)u, v).$$

Since $P^{-1}(y)$ is invertible, we get $\sigma(x,y) = \tau(x,y)$ and hence $\tau(x,y)$ is non-singular. Thus, the Jordan algebra $\mathcal{A}$ is semisimple.

If we use the notations of §1, formula (1) gives $Y \subset Y_{\mathcal{A}}$ and since $\omega(y)$ vanishes on the boundary of Y, equation (3) shows

$$Y = Y_{\mathcal{A}}, \quad \omega(y) = \omega_{\mathcal{A}}(y), \quad c = c_{\mathcal{A}}.$$

Summarizing we obtain

Theorem 6. *Given an ω–domain (Y, ω, c) there exists a real semisimple Jordan algebra $\mathcal{A}$ such that*

$$(Y, \omega, c) = \Omega(\mathcal{A}) = (Y_{\mathcal{A}}, \omega_{\mathcal{A}}, c_{\mathcal{A}}).$$

With Corollary 1 this theorem immediately leads to

Theorem 7. *Let (Y, ω, c) be an ω–domain. Then the group $\Sigma(Y, \omega)$ acts transitively on Y.*

Sometimes we denote the Jordan algebra $\mathcal{A}$ obtained by this process by

$$\mathcal{A} = \Phi(Y, \omega, c).$$

Φ defines a map from the set of ω-domains to the set of semisimple Jordan algebras on a given vector space X. Theorem 6 then says that

$$(4) \qquad\qquad \Omega \circ \Phi = \mathrm{Id}.$$

Now let us consider the map $\Phi \circ \Omega$, which maps the set of semisimple Jordan algebras on X into itself. In order to describe $\Phi \circ \Omega(\mathcal{A})$ for a semisimple Jordan algebra, we first consider the ω-domain

$$\Omega(\mathcal{A}) = (Y_A, \omega_A, c_A).$$

We know from Theorem 1 that the corresponding transformation $H(y)$ is given by

$$H(y) = P^{-1}(y),$$

where $P(y)$ is the quadratic representation of $\mathcal{A}$. We obtain the composition of the Jordan algebra $\Phi \circ \Omega(\mathcal{A})$ from the linear terms in the expansion around the point c

$$H(c + x) = P^{-1}(c + x) = (\mathrm{Id} + 2L(x) + P(x))^{-1} = \mathrm{Id} - 2L(x) + \dots .$$

Thus the products in $\mathcal{A}$ and in $\Phi \circ \Omega(\mathcal{A})$ coincide. The result is

$$\Phi \circ \Omega = \mathrm{Id}.$$

This, together with (4), yields

Theorem 8. *The map Ω from the set of semisimple Jordan algebras on a given vector space X to the set of ω-domains on X is a bijection.*

§3. Jordan algebras with equivalent ω-domains

In the definition of an ω-domain (Y, ω, c) there is some arbitrariness in the choice of the point $c \in Y$. In order to investigate how an ω-domain changes with the point c we consider two ω-domains (Y, ω, c) and (Y', ω', c') such that $Y' = Y$, $\omega' = \omega$. Since every ω-domain can be described in the form $\Omega(\mathcal{A})$ for some semisimple Jordan algebra $\mathcal{A}$ we may assume that both of them are the ω-domains of Jordan algebras.

Theorem 9. *Let $\mathcal{A}$ and $\mathcal{B}$ be semisimple Jordan algebras on the same vector space. Then we get $Y_A = Y_B$ and $\omega_A = \gamma \cdot \omega_B$ for some scalar $\gamma \neq 0$ if and only if $\mathcal{B}$ is a mutation $\mathcal{A}_f$ of $\mathcal{A}$, for some $f \in Y_A$.*

Proof. Let $\mathcal{B} = \mathcal{A}_f$ be a mutation of $\mathcal{A}$, hence $X_\mathcal{B} = X_\mathcal{A}$. Since $Y_\mathcal{B}$ is a component of the unit element f^{-1} of $\mathcal{B}$, Corollary 3 shows $f \in Y_\mathcal{B}$. Hence it follows that $Y_\mathcal{A} \cap Y_\mathcal{B} \neq \emptyset$ and $Y_\mathcal{A} = Y_\mathcal{B}$, because both sets are components of $X_\mathcal{B} = X_\mathcal{A}$.

Conversely, let $Y_\mathcal{A} = Y_\mathcal{B}$ and $\omega_\mathcal{A} = \gamma \cdot \omega_\mathcal{B}$. Hence the algebras $\mathcal{A}$ and $\mathcal{B}$ are similar and the Corollary IV.2 yields $\mathcal{B} = \mathcal{A}_f$. Since the unit element f^{-1} of $\mathcal{A}_f$ belongs to $Y_\mathcal{B} = Y_\mathcal{A}$, Corollary 3 yields $f \in Y_\mathcal{A}$. $\square$

We call two ω–domains (Y, ω, c) and (Y', ω', c') on vector spaces X resp. X' *equivalent* if there exists a bijective linear transformation $V : X \to X'$ such that

(E.1) $Y' = VY$,

(E.2) $\omega'(Vy) = \gamma \cdot \omega(y)$ for all $y \in Y$,

for some constant $\gamma \neq 0$.

Now let us consider a Jordan algebra $\mathcal{A}$ on the vector space X. Given a bijective linear transformation $V : X \to X'$, we define a Jordan algebra $V\mathcal{A} = (X', \perp)$ on X' by $x \perp y = V(V^{-1}x \circ V^{-1}y)$ for $x, y \in X'$. Obviously, $\mathcal{A}$ and $V\mathcal{A}$ are isomorphic.

Vice versa, let $\mathcal{A}'$ be a Jordan algebra on X' which is isomorphic to $\mathcal{A}$. Then there exists a bijective linear transformation V such that $\mathcal{A}' = V\mathcal{A}$.

Theorem 10. *Let $\mathcal{A}$ and $\mathcal{B}$ be semisimple Jordan algebras. The ω–domains $\Omega(\mathcal{A})$ and $\Omega(\mathcal{B})$ are equivalent if and only if $\mathcal{A}$ is isomorphic to $\mathcal{B}$.*

Proof. Let $\mathcal{A}$ and $\mathcal{A}'$ be isomorphic. Then there exists a non-singular linear transformation $V : X \to X'$ such that $\mathcal{A}' = V\mathcal{A}$. If $P(u)$ resp. $P'(u)$ are the quadratic representations of $\mathcal{A}$ resp. $\mathcal{A}'$, we get $P'(Vx) = VP(x)V^{-1}$ for $x \in X$ and therefore $\omega_{\mathcal{A}'}(Vx) = \gamma \cdot \omega_\mathcal{A}(x)$. Moreover it follows that $Y_{\mathcal{A}'} = VY_\mathcal{A}$, hence $\Omega(\mathcal{A})$ and $\Omega(\mathcal{A}')$ are equivalent.

Now let $\Omega(\mathcal{A})$ and $\Omega(\mathcal{A}')$ be equivalent, that is

$$Y_{\mathcal{A}'} = VY_\mathcal{A}, \quad \omega_{\mathcal{A}'}(Vx) = \gamma \cdot \omega_\mathcal{A}(x)$$

for some $V : X \to X'$, and some constant $\gamma \neq 0$. Consider the algebra $\tilde{\mathcal{A}} = V\mathcal{A}$ on X'. We obtain

$$Y_{\tilde{\mathcal{A}}} = VY_\mathcal{A} = Y_{\mathcal{A}'}, \quad \omega_{\tilde{\mathcal{A}}}(Vx) = \gamma_1 \cdot \omega_\mathcal{A}(x) = \gamma_2 \cdot \omega_{\mathcal{A}'}(Vx).$$

Since $\mathcal{A}$ and $\tilde{\mathcal{A}}$ are isomorphic, it therefore remains to consider the case of two Jordan algebras $\mathcal{A}$ and $\mathcal{B}$ on X such that

$$Y_\mathcal{A} = Y_\mathcal{B} \quad \text{and} \quad \omega_\mathcal{A}(x) = \gamma \cdot \omega_\mathcal{B}(x)$$

holds, and to prove that $\mathcal{A}$ and $\mathcal{B}$ are isomorphic. It follows from Theorem 9 that $\mathcal{B} = \mathcal{A}_f$ for some $f \in Y_A$. Since the subgroup $\Sigma(Y_A, \omega_A)$ of $\Gamma(\mathcal{A})$ acts transitively on Y_A, we get $\mathcal{B} = \mathcal{A}_{W_c}$ for some $W \in \Gamma(\mathcal{A})$. Now Theorem IV.7 shows that $\mathcal{B}$ and $\mathcal{A}$ are isomorphic. $\qquad \square$

Let us have another look at the components of X_A for a semisimple Jordan algebra $\mathcal{A}$ on X. We know from Theorem 3 that each connected component of X_A is an ω–domain (Y_A, ω_A, c_A), where $\mathcal{B} = \mathcal{A}_f$ is a semisimple mutation of $\mathcal{A}$. Theorem 10 states that there is a one-to-one correspondence between the equivalence classes of the connected components of X_A and the isomorphism classes of the semisimple mutations of $\mathcal{A}$.

§4. Formally real Jordan algebras

A real Jordan algebra $\mathcal{A}$ is called *formally real* if $u^2 + v^2 = 0$ implies $u = v = 0$. Obviously, in a formally real Jordan algebra 0 is the unique nilpotent element.

The elements u of any real Jordan algebra $\mathcal{A}$ have minimal decompositions in $\mathcal{A}(i)$; see Chapter V. In the formally real case, there is no need for passing to $\mathcal{A}(i)$.

Lemma 1. *Let $\mathcal{A}$ be a formally real Jordan algebra. Then the elements $u \in \mathcal{A}$ have only real eigenvalues. Given an eigenvalue of u there exists an eigenvector in $\mathbf{R}[u]$.*

Proof. Let $\xi \in \mathbf{C}$ be an eigenvalue and $z \in \mathbf{C}[u]$ be an eigenvector of u, i.e. $u \circ z = \xi z$. Since u is real, we get $u \circ \bar{z} = \bar{\xi}\, \bar{z}$ and $\bar{z}, z$ commute. Now it follows that

$$\xi z \circ \bar{z} = (u \circ z) \circ \bar{z} = (u \circ \bar{z}) \circ z = \bar{\xi} z \circ \bar{z}.$$

With $z = x + iy$, $x, y \in \mathbf{R}[u]$, it follows that $z \circ \bar{z} = x^2 + y^2 \neq 0$, and hence $\bar{\xi} = \xi$. Since

$$\xi(x + iy) = \xi z = u \circ z = u \circ x + iu \circ y,$$

it is clear that x and y are eigenvectors of u. $\qquad \square$

The Remark in Chapter V, §3 in combination with Theorems V.2 and V.3 yields

Theorem 11. *Let $\mathcal{A}$ be a formally real Jordan algebra, and $u \in \mathcal{A}$. Then the roots of the minimal polynomial of u are real and simple, and u has a minimal decomposition*

$$u = \sum_k \xi_k c_k, \quad c_k \in \mathbf{R}[u].$$

The minimal decomposition is uniquely determined up to permutation of the idempotents.

In order to obtain a criterion to decide whether a given Jordan algebra is formally real, we prove

Theorem 12. *Let A be a real Jordan algebra. Then the following statements are equivalent:*

(A) A is formally real.

(B) The bilinear form $\tau(x,y) = \mathrm{tr}L(x \circ y)$ is positive definite.

(C) There exists a positive definite bilinear form $\sigma(x,y)$ satisfying

$$\sigma(x, y \circ z) = \sigma(x \circ y, z).$$

Corollary 5. *A formally real Jordan algebra is semisimple.*

Proof. "(A) $\Rightarrow$ (B)": It follows from Lemma 1 that $x \in A$ has only real eigenvalues ξ_k. Hence x^2 has the eigenvalues ξ_k^2. Theorem V.5 then shows that $L(x^2)$ has the ξ_k^2 and some of the numbers $\frac{1}{2}\left(\xi_k^2 + \xi_l^2\right)$ as eigenvalues. We get $\tau(x,x) = \mathrm{tr}\,L(x \circ x) \geq 0$. Moreover, $\tau(x,x) = 0$ means that all the eigenvalues of x are zero, and Theorem 11 yields $x = 0$.

"(B) $\Rightarrow$ (C)": Take $\sigma(x,y) = \tau(x,y)$ and use Lemma III.4.

"(C) $\Rightarrow$ (A)": Let $x, y \in A$ such that $x^2 + y^2 = 0$. It follows that

$$\sigma(x,x) + \sigma(y,y) = \sigma(c,x^2) + \sigma(c,y^2) = 0.$$

Since σ is positive definite, we obtain $x = y = 0$. $\square$

Given a self-adjoint linear transformation A (with respect to a bilinear form $\tau(x,y)$) the notions *positive definite* $(A > 0)$ resp. *positive semidefinite* $(A \geq 0)$ were defined in Chapter I, §1. If $\tau(x,y)$ is a positive definite bilinear form, a given transformation $A = A^*$ is positive definite resp. positive semidefinite if and only if the eigenvalues of A are positive resp. non-negative.

Theorem 13. *Given a formally real Jordan algebra A one has the following descriptions of the ω–domain of A:*

$$Y_A = \{y \in A;\ L(y) > 0\} = \{y \in A;\ \text{the eigenvalues of } y \text{ are positive}\}$$
$$= \{e(x);\ x \in A\} = \{x^2;\ x \in A, \det P(x) \neq 0\},$$
$$\overline{Y}_A = \{y \in A;\ L(y) \geq 0\} = \{y \in A;\ \text{the eigenvalues of } y \text{ are non-negative}\}$$
$$= \{x^2;\ x \in A\}.$$

Proof. Let $Y = \{y \in \mathcal{A}; L(y) > 0\}$. Then Y is an open and convex, hence a connected set. Since $\det L(y) \neq 0$ implies $\det P(y) \neq 0$, we get $Y \subset Y_{\mathcal{A}}$. If Y were a proper subset of $Y_{\mathcal{A}}$ then some point $a \in Y_{\mathcal{A}}$ would belong to the boundary of Y. Hence $L(a) \geq 0$, and $L(a)$ is not positive definite. This implies $\det L(a) = 0$. If α_k are the eigenvalues of a we know from Theorem V.5 that $L(a)$ has the α_k and some of the $\frac{1}{2}(\alpha_k + \alpha_l)$ as eigenvalues. These numbers are non-negative and at least one of them is zero. Hence at least one eigenvalue of a is zero and we get a contradiction to $a \in Y_{\mathcal{A}}$. Therefore $Y = Y_{\mathcal{A}}$. The next description of $Y_{\mathcal{A}}$ also follows from Theorem V.5. Theorem 4 shows that the third and the fourth set are equal, and we only need to prove that every $y \in Y_{\mathcal{A}}$ is a square. This is obvious, since the eigenvalues of y are positive. The description of $\overline{Y}_{\mathcal{A}}$ follows in a similar way. $\qquad\square$

Since the eigenvalues of the elements of $Y_{\mathcal{A}}$ are positive, we see immediately that the map $x \mapsto x^2$ is a bijection of $Y_{\mathcal{A}}$ (resp. $\overline{Y}_{\mathcal{A}}$) onto itself.

§5. Homogeneous domains of positivity

We recall the notion of a domain of positivity from Chapter I. Let $\sigma(x, y)$ be a positive definite bilinear form on the vector space X. Then $Y \subset X$ is called a *domain of positivity* (with respect to σ) if Y is open and

(P.1) $a, b \in Y$ implies $\sigma(a, b) > 0$.

(P.2) $x \in X$ and $\sigma(a, x) > 0$ for every $a \in \overline{Y}$, $a \neq 0$, implies $x \in Y$.

$\Sigma(Y)$ denotes the group of all linear transformations $W : X \to X$ such that $WY = Y$. Moreover, Y is called *homogeneous* if $\Sigma(Y)$ acts transitively on Y. A *norm* $\omega(y)$ of Y satisfies, by definition, $\omega(Wy) = |\det W| \cdot \omega(y)$ for $W \in \Sigma(Y)$ and $y \in Y$. For homogeneous domains of positivity there is only one norm, up to a constant factor.

Theorem 14. *The ω–domain $Y_{\mathcal{A}} = \Omega(\mathcal{A})$ of a formally real Jordan algebra $\mathcal{A}$ is a homogeneous domain of positivity (with respect to the bilinear form τ on $\mathcal{A}$) with norm $\omega(y) = \gamma \cdot \sqrt{|\det P(y)|}$. The automorphism group $\Sigma(Y_{\mathcal{A}})$ is a subgroup of the structure group $\Gamma(\mathcal{A})$.*

Proof. Let $\Omega(\mathcal{A}) = (Y_{\mathcal{A}}, \omega_{\mathcal{A}}, c_{\mathcal{A}})$ and $a, b \in Y_{\mathcal{A}}$. Theorem 13 states that there is an element $x \in \mathcal{A}$, $\det P(x) \neq 0$, such that $a = x^2$. Then $\tau(x, y) = L(x \circ y)$ satisfies

$$\tau(a, b) = \tau(x^2, b) = \tau(L(b)x, x) > 0$$

because of $L(b) > 0$. This proves (P.1). In order to demonstrate (P.2) let $x \in A$ such that $\tau(a, x) > 0$ for every $a \in \overline{Y}_A, a \neq 0$. We put $a = u^2, u \neq 0$, and we get

$$\tau(a, x) = \tau(u^2, x) = \tau(L(x)u, u) > 0.$$

Since $u \neq 0$ is arbitrary, it follows that $L(x) > 0$, and $x \in Y_A$.

As $\Sigma(Y_A, \omega_A)$ is a subgroup of $\Sigma(Y_A)$, Corollary 1 shows that $\Sigma(Y_A)$ acts transitively on Y_A, i.e. Y_A is homogeneous. Hence the norm of Y_A is uniquely determined up to a constant. Since $\sqrt{|\det P(y)|}$ is a norm, we get $\omega(y) = \gamma \cdot \sqrt{|\det P(y)|}$. Moreover, if ω_0 denotes the special norm from Theorem I.4 then $\omega_0(Wy) = |\det W| \cdot \omega_0(y)$ for all $W \in \Sigma(Y_A)$, $y \in Y_A$ by equation (1) in Chapter I, §4. This shows $\Sigma(Y_A) = \Sigma(Y_A, \omega_A) \subset \Gamma(A)$. $\square$

Vice versa, let us consider a given homogeneous domain of positivity Y (with respect to the positive definite bilinear form $\sigma(x, y)$). We know from Theorem II.2 that (Y, ω, c) is an ω–domain. Then $\omega = \omega(y)$ is a norm on Y, and c is the fixed point of the involution of Y. Moreover, the bilinear form σ is equal to the bilinear form associated with the ω–domain; i.e., obtained from the second logarithmic derivative of ω at c. Let A be the uniquely determined Jordan algebra such that $(Y, \omega, c) = \Omega(A)$. Theorem II.4 shows that

$$\sigma(x \circ y, z) = \sigma(x, y \circ z).$$

Since σ is positive definite, Theorem 12 implies that A is formally real. In combination with Theorem 14 we have proved

Theorem 15. *The map $A \mapsto \Omega(A)$ is a bijection from the set of formally real Jordan algebras on a given vector space X onto the set of homogeneous domains of positivity (with respect to a positive definite bilinear form) on X.*

This theorem gives an algebraic characterization of those semisimple Jordan algebras which are associated with domains of positivity. There is also a geometric characterization of the domains of positivity in the family of ω–domains:

Theorem 16. *An ω–domain (Y, ω, c) is a domain of positivity if and only if the set Y is convex.*

Proof. Since every domain of positivity is a convex cone, we only need to show that Y is a domain of positivity if Y is convex. Let $y \in Y$ and $(Y, \omega, c) = \Omega(A)$, with A a semisimple Jordan algebra. We conclude from Theorem 4 that for $m = 0, 1, \ldots$ the powers y^{2^m} belong to Y. Le ξ be an eigenvalue of y, then ξ^{2^m} is an eigenvalue of y^{2^m}. Let us assume that y has an eigenvalue

ξ that is not real and positive. Then there is an m such that the real part of $\eta = \xi^{2^m}$ is negative. Hence we get

$$\eta^2 + \alpha\eta + \beta = 0, \quad \alpha = -2\mathrm{Re}\,(\eta)\,, \ \beta = |\eta|,$$

with positive coefficients. Hence zero is an eigenvalue of

$$y^{2^{m+1}} + \alpha y^{2^m} + \beta c.$$

Since Y is convex, we get a contradiction. Therefore the eigenvalues of $y \in Y$ are positive, and the eigenvalues of $y \in \overline{Y}$ are non-negative.

Now we consider $\tau(x,x) = \mathrm{tr}\,L(x^2)$ for $x \in \mathcal{A}$. In view of $x^2 \in \overline{Y}$ the eigenvalues of x^2 are non-negative, and Theorem V.5 states that $\tau(x,x) \geq 0$. Since $\tau(x,y)$ is non-singular, it follows that $\tau(x,y)$ is positive definite. Theorem 12 implies that $\mathcal{A}$ is formally real. Hence $\Omega(\mathcal{A})$ is a domain of positivity.   □

Now let us consider the geodesics of a domain of positivity Y. By the definition of geodesics of an ω–domain in Chapter II, §3, the geodesics are the solutions of the variational problem relative to $\lambda(y,x) = \sigma(H(y)x,x)$, that is they are the solutions of the Euler–Lagrange differential equation

$$(1) \qquad \frac{\partial\lambda}{\partial y} - \frac{d}{d\tau}\frac{\partial\lambda}{\partial\dot{y}} = 0, \quad \lambda = \lambda(y,\dot{y}).$$

Note that in the case of domains of positivity the bilinear form σ and the linear transformations $H(y)$ are positive definite, hence $\lambda(y,x) \geq 0$ for all $y \in Y$ and $x \in X$. Let $y = y(\tau)$ be a solution of (1), and consider $\lambda(\tau) := \lambda(y(\tau), \dot{y}(\tau))$. Then one gets

$$\dot{\lambda} = \frac{\partial\lambda}{\partial\dot{y}}\ddot{y} + \frac{\partial\lambda}{\partial y}\dot{y} = \frac{\partial\lambda}{\partial\dot{y}}\ddot{y} + \left(\frac{d}{d\tau}\frac{\partial\lambda}{\partial\dot{y}}\right)\dot{y} = \frac{d}{d\tau}\left(\frac{\partial\lambda}{\partial\dot{y}}\dot{y}\right).$$

Since λ is homogeneous of degree 2, the right hand side is equal to $2\dot{\lambda}$, and we obtain $\dot{\lambda} = 0$. Hence $\lambda(y(\tau), \dot{y}(\tau))$ is constant.

Using this, a routine computation shows that every (nonconstant) solution of (1) also solves the Euler–Lagrange equation with respect to the functional

$$\mu(y,x) = \sqrt{\lambda(y,x)}.$$

The extremals of the variational problem corresponding to μ are the "geometric geodesics" of Y (see also Chapter I, §7). As we have just seen, these two notions coincide in domains of positivity.

We know from Theorems II.3 and II.4 that the curves $y(\tau) = e(\tau x)$ are exactly the geodesics through the point c. Moreover, τ is the arc length provided that $\sigma(x,x) = 1$ holds. Theorem 13 states that the map $X \to Y$, $u \mapsto e(u)$, is a surjection. Therefore, given a point $a \in Y$, there exists a geodesic through c and a. Moreover, we get

Theorem 17. *In a homogeneous domain of positivity any two points can be joined by one and only one geodesic.*

Proof. Since Y is homogeneous, we only need to consider the geodesics through c and to prove that $e(a) = e(b)$ implies $a = b$. In order to do so, we use the minimal decompositions from Theorem 11:

$$a = \sum_k \alpha_k c_k, \quad b = \sum_k \beta_k d_k,$$

where the eigenvalues α_k resp. β_k and the idempotents c_k resp. d_k are real. It follows that

$$\sum_k e^{\alpha_k} c_k = \sum_k e^{\beta_k} d_k,$$

and uniqueness shows that we may asssume $c_k = d_k$ and $e^{\alpha_k} = e^{\beta_k}$ for all k. Now $\alpha_k = \beta_k$ follows for all k. See also the next section for another proof. □

§6. Elementary functions on formally real Jordan algebras

Let $\mathcal{A}$ be a formally real Jordan algebra over $\mathbf{R}$ with underlying vector space X. Consider a power series

$$f(\tau) = \sum_{k=0}^{\infty} \alpha_k \tau^k, \quad \text{and let} \quad s_m(\tau) = \sum_{k=0}^{m} \alpha_k \tau^k.$$

We set

$$f(u) = \sum_{k=0}^{\infty} \alpha_k u^k,$$

whenever the series is convergent. We will prove that this series is convergent if and only if the series $f(\xi_\nu)$ are convergent for all the eigenvalues ξ_ν of u. Let

$$u = \sum_\nu \xi_\nu c_\nu$$

be the minimal decomposition of u. Then we get

$$s_m(u) = \sum_\nu s_m(\xi_\nu) c_\nu.$$

Multiplying this equation by c_ν we see that $f(u)$ is convergent if and only if the sequence $s_m(\xi_\nu)$ converges as $m \to \infty$. In addition,

$$(1) \qquad f(u) = \sum_\nu f(\xi_\nu)c_\nu.$$

Now let $f(\tau)$ be an arbitrary function defined on a subset $M \subset \mathbf{R}$. We define $f(u)$ by formula (1) if every eigenvalue of u belongs to M.

Obviously we get $f(\tau) = g(\tau) + h(\tau)$, resp. $f(\tau) = g(\tau)h(\tau)$, resp. $f(\tau) = g(h(\tau))$ implies $f(u) = g(u) + h(u)$, resp. $f(u) = g(u) \circ h(u)$, resp. $f(u) = g(h(u))$.

The domain of positivity corresponding to the algebra $\mathcal{A}$ gives rise to the orderings ">" and "$\geq$" on the vector space underlying $\mathcal{A}$ (see Chapter I, §2). If $f(\tau)$ is defined on the interval $\alpha < \tau < \beta$ then $f(u)$ is defined if all the eigenvalues of u belong to this interval. From

$$u - \alpha c = \sum_\nu (\xi_\nu - \alpha)c_\nu \ , \ \beta c - u = \sum_\nu (\beta - \xi_\nu)c_\nu$$

it follows that $f(u)$ is defined if and only if $\alpha c < u < \beta c$. This describes an open bounded subset of X. In the same way we find that $f(u)$ is defined on

$$\{u \in X; \ \alpha c \leq u \leq \beta c\}$$

if $f(\tau)$ is defined for $\alpha \leq \tau \leq \beta$. An inequality for the values of $f(\tau)$ implies the corresponding inequality for $f(u)$. For instance, $\gamma \leq f(\tau) \leq \delta$ yields $\gamma c \leq f(u) \leq \delta c$.

Let $f(\tau)$ be a continuous and monotone function defined on the inverval $\alpha \leq \tau \leq \beta$. Then the inverse function $g(\tau)$ is defined on an interval $\gamma \leq \tau \leq \delta$ and we get $f(g(\tau)) = \tau$ for $\gamma \leq \tau \leq \delta$ resp. $g(f(\tau)) = \tau$ for $\alpha \leq \tau \leq \beta$. Hence $f(u)$ resp. $g(v)$ are defined on $\alpha c \leq u \leq \beta c$ resp. $\gamma c \leq v \leq \delta c$ and we obtain

$$f(g(v)) = v \quad \text{resp.} \quad g(f(u)) = u.$$

Therefore, $u \mapsto f(u)$ is a bijective map from the set $\{u \in X; \ \alpha c \leq u \leq \beta c\}$ onto the set $\{v \in X; \ \gamma c \leq v \leq \delta c\}$.

We can apply this result to the function e^τ and see once more that the map $X \to Y$, $u \mapsto e(u)$, is a bijection.

For another application we consider the function

$$f(\tau) = (1 - \tau)(1 + \tau)^{-1} = \frac{2}{1 + \tau} - 1.$$

$\tau \mapsto f(\tau)$ is a bijection from the interval $]0; \infty[$ onto the interval $] - 1; 1[$. Hence the function

$$f(u) = (c - u) \circ (c + u)^{-1}$$

maps Y bijectively onto the domain $\{v \in X;\ -c < v < c\}$. With the above argument we conclude

(2) $$u \in X, \quad f(u) \in Y \quad \Leftrightarrow \quad -c < u < c.$$

§7. Direct sums

We consider a semisimple Jordan algebra $\mathcal{A}$ on X, which is a direct sum of two subalgebras: $\mathcal{A} = \mathcal{A}_1 \oplus \mathcal{A}_2$. From the definition of a direct sum (Chapter III, §4) it follows that the underlying vector space X is a direct sum of vector spaces X_1 and X_2 such that $\mathcal{A}_1$ resp. $\mathcal{A}_2$ are Jordan algebras on X_1 resp. X_2 and furthermore $\mathcal{A}_1 \circ \mathcal{A}_2 = \{0\}$. In the notation of Chapter IV, §6, the quadratic representation is given by

$$P(x) = P(x_1) \times P(x_2), \quad \text{with} \quad x = x_1 + x_2, \quad x_i \in \mathcal{A}_i,$$

and also $L(x) = L(x_1) \times L(x_2)$. This implies

$$\tau(x, y) = \tau_1(x_1, y_1) + \tau_2(x_2, y_2),$$

with τ, τ_1 and τ_2 denoting the bilinear forms associated with $\mathcal{A}$, $\mathcal{A}_1$ resp. $\mathcal{A}_2$. Obviously, the ω–domains are given by

(1) $$Y_\mathcal{A} = Y_{\mathcal{A}_1} + Y_{\mathcal{A}_2} = \{y_1 + y_2;\ y_i \in Y_{\mathcal{A}_i}\},$$
$$\omega_\mathcal{A}(y) = \omega_{\mathcal{A}_1}(y_1) \cdot \omega_{\mathcal{A}_2}(y_2),$$
$$c_\mathcal{A} = c_{\mathcal{A}_1} + c_{\mathcal{A}_2}.$$

Conversely, let us consider the two ω–domains (Y_1, ω_1, c_1) on X_1 resp. (Y_2, ω_2, c_2) on X_2. Form the triple

(2) $$(Y, \omega, c) = (Y_1, \omega_1, c_1) + (Y_2, \omega_2, c_2),$$

where the subset Y of the direct sum $X_1 + X_2$ is defined by $Y = Y_1 + Y_2$, and where $\omega(y) = \omega_1(y_1) \cdot \omega_2(y_2)$, $c = c_1 + c_2$. Since the bilinear form associated with (Y, ω, c) is the sum of the bilinear forms associated with both given ω–domains, it follows that the sum (2) is an ω–domain on X. Using formula (1), and this notation, we get for semisimple Jordan algebras $\mathcal{A}_i$:

(3) $$\Omega(\mathcal{A}_1 \oplus \ldots \oplus \mathcal{A}_r) = \Omega(\mathcal{A}_1) + \ldots + \Omega(\mathcal{A}_r).$$

As an abbreviation let $\Sigma(\mathcal{A}) = \Sigma(Y_\mathcal{A}, \omega_\mathcal{A})$ be the group of automorphisms of the ω–domain $\Omega(\mathcal{A})$, with $\mathcal{A}$ semisimple. We know that

$$\Sigma(A) = \{W \in \Gamma(A); \ WY_A = Y\}.$$

Let

$$A = A_1 \oplus \ldots \oplus A_r$$

be the decomposition of A into simple Jordan algebras A_i. We know from Theorem IV.10 that the structure group satisfies

$$\Gamma(A) = \Gamma_0(A) \cdot \Gamma_1(A),$$

where $\Gamma_0(A)$ is a finite group of certain permutations of the A_i, and

$$\Gamma_1(A) = \Gamma(A_1) \times \ldots \times \Gamma(A_r).$$

If A_i and A_j are isomorphic then their ω–domains are equivalent with respect to the linear transformation defining the isomorphism. Therefore

$$\Gamma_0(A) \subset \Sigma(A).$$

Obviously, $\Sigma(A_1) \times \ldots \times \Sigma(A_r) \subset \Sigma(A)$. On the other hand let

$$W = W_1 \times \ldots \times W_r \in \Gamma_1(A)$$

be an element in $\Sigma(A)$. Then W_i is a linear transformation of the vector space X_i underlying A_i, and $WY_A = Y_A$ implies $W_i Y_{A_i} = Y_{A_i}$. Therefore $W \in \Sigma_1(A)$, where $\Sigma_1(A)$ stands for $\Sigma(A_1) \times \ldots \times \Sigma(A_r)$. Summarizing we obtain

Theorem 18. *Given a semisimple Jordan algebra A we have*

$$\Sigma(A) = \Gamma_0(A) \cdot \Sigma_1(A).$$

Notes

Ch. Hertneck proved in [4] that there is a domain of positivity associated with a formally real Jordan algebra as we described in section 5. For domains of positivity we follow [0].

Editors' notes

1. (i) In the proof of Theorem 1 the following topological fact is used: If U and V are open connected subsets of $\mathbf{R}^n$ with $U \cap V \neq \emptyset$ and $U \cap \partial V = \emptyset$, then $U \subset V$.

(ii) To prove injectivity of $x \mapsto x^2$ on $\overline{Y_A}$ in the case of a formally real Jordan algebra (as asserted at the end of §4), one may argue as in the proof of Theorem 17: Consider $x, y \in \overline{Y_A}$, with minimal decompositions $x = \sum_k \alpha_k c_k$, $y = \sum_k \beta_k d_k$, with non-negative α_k and β_k. Then $x^2 = y^2$ implies that $\sum_k \alpha_k^2 c_k = \sum_k \beta_k^2 d_k$, and these two minimal decompositions, due to uniqueness, must coincide. Thus, after possible relabeling, one has $c_k = d_k$ and $\alpha_k = \beta_k$ for all k.

(iii) The scalar products $\sigma(H(y)x, x)$ endow the domain of positivity Y_A with the structure of a Riemannian manifold. See also Chapter I, §7.

(iv) It is a well-known fact that geodesics of a Riemannian manifold are extremals of both the length functional μ and the energy functional λ. See Jost's book [Jos], Chapter 1, Section 1.4. In some cases (e.g. in dimension one) the Euler–Lagrange equations for μ become degenerate and yield $0 = 0$.

(v) Formally real Jordan algebras are nowadays often called *Euclidean* Jordan algebras.

2. The monograph [FKo] by Faraut and Korányi contains a well-written exposition of the correspondence between formally real Jordan algebras and domains of positivity which does not (directly) address the broader concept of ω–domains. The underlying machinery of Koecher's approach is also outlined in [KrP], and in Petersson's contribution to [KMP]. The correspondence between formally real Jordan algebras and self-dual homogeneous cones is also described by Satake [Sa2], Chap. I. Concerning applications of Jordan algebras in differential geometry we mention Helwig [H2], U. Hirzebruch [Hir], and the fundamental book [Lo1] by Loos.

Chapter VII. Half-Spaces

§1. The half-space of a semisimple Jordan algebra

In this chapter let A be a semisimple Jordan algebra of the finite dimension n over the field $\mathbf{R}$ of real numbers. Let c be the unit element of A and let $L(x)$ resp. $P(x)$ be the linear transformations associated with A. If X is the vector space underlying A, then we consider on $X(i)$ the complex extension $A(i)$ of A (see Chapter V, §6). Given $z = x + iy \in X(i)$, $x, y \in X$, we define the *real* and *imaginary part* of z by

$$\operatorname{Re}(z) := x, \quad \operatorname{Im}(z) := y.$$

If $b_1, \dots, b_n$ is a basis of A then $X(i)$ contains exactly the points $\xi_1 b_1 + \dots + \xi_n b_n$, where $\xi_1, \dots, \xi_n$ are arbitrary complex numbers. The elements $b_1, \dots, b_n$ form a basis of $X(i)$ over $\mathbf{C}$. We will consider only such bases of $X(i)$.

Let f be a scalar- or vector-valued function defined on an open subset of $X(i)$. Then $f(z)$ is called *holomorphic* on this set if $f(z)$ itself resp. all the components of $f(z)$ with respect to the basis $b_1, \dots, b_n$ are holomorphic functions in the components of z. Obviously, this property is independent of the choice of the basis. Analogously, we have the notion of a map $z \mapsto A(z)$ into the set of linear transformations on $X(i)$ to be holomorphic. Moreover, we consider the ω–domain $\Omega(A) = (Y, \omega, c)$ and the group

$$\Sigma = \{ W \in \Gamma(A); \, WY = Y \}.$$

We know that Σ acts transitively on Y.

Given a complex 2×2 matrix M with non-vanishing determinant, we define a generalized *Möbius transformation*, namely a map $z \mapsto Mz$ on $X(i)$ by

$$Mz := (\alpha z + \beta c) \circ (\gamma z + \delta c)^{-1} \quad \text{if} \quad M = \begin{pmatrix} \alpha & \beta \\ \gamma & \delta \end{pmatrix}.$$

Mz is defined if and only if the inverse of $\gamma z + \delta c$ exists, i.e. $\det P(\gamma z + \delta c) \neq 0$. This shows that Mz is holomorphic on an open, connected, and dense subset of $X(i)$. Moreover, if the element Mz is defined, then Mz belongs to $\mathbf{C}[z]$.

In the same way as for the complex plane we get for M_1, $M_2 \in GL(2, \mathbf{C})$:

$$(M_1 M_2)z = M_1(M_2 z),$$

$$M_1 z = M_2 z \quad \text{for every} \quad z \quad \Leftrightarrow \quad M_1 = \lambda M_2 \quad \text{for some} \quad \lambda \in \mathbf{C}.$$

Thus the group of generalized Möbius transformations is isomorphic to $PSL(2, \mathbf{C})$. Moreover one has for $M = \begin{pmatrix} \alpha & \beta \\ \gamma & \delta \end{pmatrix} \in GL(2, \mathbf{C})$.

$$(1) \qquad Mz = \begin{cases} \frac{\alpha}{\delta}z + \frac{\beta}{\delta}c & \text{if} \quad \gamma = 0, \\ \frac{\alpha}{\gamma}c - \frac{\det M}{\gamma}(\gamma z + \delta c)^{-1} & \text{if} \quad \gamma \neq 0. \end{cases}$$

The bilinear form $\tau(z, w) = \operatorname{tr} L(z \circ w)$ is non-singular and the linear transformations $L(z)$ and $P(z)$ are self-adjoint with respect to τ. Using the formula

$$\frac{\partial z^{-1}}{\partial z} = -P^{-1}(z)$$

we can compute the Jacobian of the map $z \mapsto Mz$:

$$(2) \qquad \frac{\partial Mz}{\partial z} = \det M \cdot P^{-1}(\gamma z + \delta c), \quad \text{with} \quad M = \begin{pmatrix} \alpha & \beta \\ \gamma & \delta \end{pmatrix} \in GL(2, \mathbf{C}).$$

Indeed, we may assume that $\gamma \neq 0$ holds; then this equation follows from (1). Moreover we have, by the chain rule,

$$(3) \qquad \frac{\partial M_1 M_2 z}{\partial z} = \frac{\partial M_1(M_2 z)}{\partial M_2 z} \cdot \frac{\partial M_2 z}{\partial z}.$$

Using the ω–domain Y of $\mathcal{A}$ we define a subset Z of $X(i)$ by

$$Z := \{z = x + iy; \; x \in X, \; y \in Y\} = X + iY.$$

Z is an open and connected subset of $\mathcal{A}(i)$, which is called the *half-space associated with the Jordan algebra $\mathcal{A}$*.

A map f from some open subset of $X(i)$ onto some other open subset of $X(i)$ is called *biholomorphic* if f is a bijective map and if f as well as f^{-1} are holomorphic. It is known that f is biholomorphic if f is a topological and holomorphic map. The Jacobian of a biholomorphic map does not vanish on the domain of definition.

Given $z \in X(i)$ we define

$$\beta(z) = \det P(z - \bar{z}).$$

Then $\beta(z)$ is a polynomial in the real and imaginary part of z. In order to get a description of the group of biholomorphic automorphisms of Z, we consider the following *set H* of biholomorphic maps: $f \in H$ if and only if f is defined on some open subset of Z, maps this subset into Z, and satisfies

(4)
$$\beta(f(z)) = \left| \det \left(\frac{\partial f(z)}{\partial z} \right) \right|^2 \cdot \beta(z),$$

whenever $f(z)$ is defined.

The identity (4) can be recast in terms of the quadratic representation, using the linear transformation

$$B(z, w) = -P(z - \overline{w}),$$

which is a polynomial in z and $\overline{w}$. We use the following abbreviation: For $z, w \in X(i)$, let $A(z, w) : X(i) \to X(i)$ be a linear transformation depending holomorphically on z and $\overline{w}$. Then for a biholomorphic map f define the linear transformation A_f by

$$A_f(z, w) = \left(\frac{\partial f(z)}{\partial z} \right)^{-1} A(f(z), f(w)) \overline{\left(\frac{\partial f(w)}{\partial w} \right)}^{*-1}.$$

If $z \mapsto g(z)$ is another map, then we get for the composite map fg (if it is defined)

(5)
$$A_{fg} = (A_f)_g.$$

Lemma 1. *Identity (4) holds for every z if and only if $B_f = B$.*

Proof. If $B_f = B$ then we obtain (4) by evaluating the determinant of B for $w = z$. Now let us assume (4) and consider $\beta(z, w) = \det P(z - \overline{w})$. The function

$$\alpha(z, w) = \beta(f(z), f(w)) - \det \left(\frac{\partial f(z)}{\partial z} \right) \cdot \det \overline{\left(\frac{\partial f(w)}{\partial w} \right)} \cdot \beta(z, w)$$

is holomorphic in z and $\overline{w}$, with z, w in the domain of definition of f, and satisfies $\alpha(z, z) = 0$; hence $\alpha(z, w) = 0$. From Lemma IV.1 it is known that

$$\frac{\partial}{\partial z} \left(\frac{\partial}{\partial \overline{w}} \log \beta(z, w) \right)^* = 2P^{-1}(z - \overline{w}),$$

and this yields $B_f = B$. □

Thus, a biholomorphic map f from an open subset of Z into Z belongs to H if and only if $B_f = B$. In order to investigate the set H we consider two subsets:

$$H^\infty = \{f \in H; \; f(z) \text{ is linear in } z\},$$

that is the set of $f \in H$ of the form $f(z) = Wz + a$, resp.

$$H_{ic} = \{f \in H; \; f(ic) = ic\},$$

i.e. f is biholomorphic in a neighborhood of ic and maps ic to ic. It follows from (5) that H_{ic} may be viewed as a group.

Lemma 2. (a) An element $f \in H$ belongs to H^∞ if and only if $f(z) = Wz + a$, with $W \in \Sigma$ and $a \in X$.

(b) H^∞ is a group of biholomorphic mappings of Z onto itself and acts transitively on Z.

(c) $H = H^\infty \cdot H_{ic} \cdot H^\infty$.

Proof. (a) If $f(z) = Wz + a$, with $W \in \Sigma$ and $a \in X$, we obtain

$$B(f(z), f(w)) = -P(W(z - \overline{w})) = WB(z, w)W^*$$

from the fact that $\Sigma \subset \Gamma(A)$. Hence $B_f = B$ follows. On the other hand, for $f(z) = Wz + a$, with W and a complex, the identity $B_f = B$ implies

$$(6) \qquad P(Wz - \overline{W}\overline{w} + b) = WP(z - \overline{w})\overline{W}^*, \quad b = a - \overline{a}.$$

Since this is a polynomial identity, it holds for all z, w. We set $w = 0$ and compare the linear terms in z:

$$L(Wz)L(b) + L(b)L(Wz) = L(b \circ Wz).$$

For $Wz = c$ it follows that $L(b) = 0$, thus $b = 0$ and a is real. Now we set $z = u + W^{-1}c$, $w = \overline{u}$ and get $P(c + (W - \overline{W})u) = \text{constant}$. Comparing the linear terms in u we find $L((W - \overline{W})u) = 0$. Hence W is real. In (6) we put $w = 0$ and obtain $W \in \Gamma(A)$. Since $z \mapsto f(z)$ maps an open subset of Z into Z, the map $y \mapsto Wy$ maps an open subset of Y into Y, hence $WY \cap Y \neq \emptyset$. This implies $WY \subset Y$, since each element of WY is invertible and Y is a connected component within the set of invertible elements of X. Analogously $W^{-1}Y \subset Y$ follows and hence $WY = Y$.

(b) Part (a) states that every $f \in H^\infty$ is a biholomorphic map of Z onto itself. Given $z = x + iy$ there is an $f \in H^\infty$ which maps z onto ic. Indeed, take an element $W \in \Sigma$ such that $Wy = c$ and put $a = -Wx$.

(c) The statement follows from (b) and (5).   □

The generalized Möbius transformations are of particular interest.

Lemma 3. *For every $M \in GL(2, \mathbf{R})$ with $\det M > 0$, the map $z \mapsto Mz$ belongs to H. In particular $z \mapsto j(z) = -z^{-1}$ is an element of H.*

Proof. A simple calculation shows that $z \mapsto Mz$ maps ic to a point in Z. From equation (1) it follows that $z \mapsto Mz$ is a composition of mappings $z \mapsto \alpha z + \beta c$ and $z \mapsto j(z)$. Since $B_f = B$ holds for $f(z) = \alpha z + \beta c$, $\alpha \neq 0$, we only need to prove that $B_j = B$. We obtain

$$B_j(z, w) = \left(\frac{\partial j(z)}{\partial z}\right)^{-1} B(j(z), j(w)) \left(\frac{\partial j(w)}{\partial w}\right)^{*-1}$$

$$= -P(z)P(z^{-1} - \overline{w}^{-1})P(\overline{w}).$$

In Theorem V.4 we derived

(7) $\qquad P(u)P(u^{-1} - v^{-1})P(v) = P(u - v) \quad \text{if} \quad \det(P(u)P(v)) \neq 0,$

and therefore $B_j = B$ follows. $\qquad\qquad\qquad\qquad\qquad\qquad\qquad\qquad\square$

§2. The isotropy group $\tilde{H}_0$

We know from Lemma 2 that the set H of mappings is characterized by the group H^∞ and the isotropy group H_{ic} which contains all biholomorphic maps f from a neighborhood of ic onto a neighborhood of ic satisfying $B_f = B$. In order to investigate H_{ic} we consider two generalized Möbius transformations, namely

(1) $\quad p(z) = (z - ic) \circ (z + ic)^{-1} = c - 2i(z + ic)^{-1}, \quad \dfrac{\partial p(z)}{\partial z} = 2iP^{-1}(z + ic),$

(2) $\quad \tilde{p}(z) = i(c + z) \circ (c - z)^{-1} = -ic + 2i(c - z)^{-1}, \quad \dfrac{\partial \tilde{p}(z)}{\partial z} = 2iP^{-1}(c - z),$

where the corresponding matrices are given by

$$M = \begin{pmatrix} 1 & -i \\ 1 & i \end{pmatrix} \quad \text{resp.} \quad \tilde{M} = \begin{pmatrix} i & i \\ -1 & 1 \end{pmatrix} = 2iM^{-1}.$$

Obviously, the mappings p and $\tilde{p}$ are inverse to each other; p is a biholomorphic map from a neighborhood of ic onto a neighborhood of 0.

We use the abbreviations

(3) $\qquad\qquad \tilde{H} = p \cdot H \cdot \tilde{p}, \quad \tilde{H}_0 = p \cdot H_{ic} \cdot \tilde{p}, \quad \tilde{B} = B_{\tilde{p}}.$

$\tilde{B}(z, w)$ is holomorphic in z and $\overline{w}$ in a neighborhood of 0. Every g in $\tilde{H}_0$ is a biholomorphic map from a neighborhood of 0 onto a neighborhood of 0; and such a map belongs to $\tilde{H}_0$ if and only if

$$\tilde{B}_g = \tilde{B},$$

because $B_f = B$ is equivalent to $\tilde{B}_{p \cdot f \cdot \tilde{p}} = \tilde{B}$.

In order to investigate the isotropy group $\tilde{H}_0$ we need to derive another formula for $\tilde{B}$:

Lemma 4. *For $\tilde{B} = B_{\tilde{p}}$ one has*

$$\tilde{B}(z, w) = \mathrm{Id} - 2L(z)L(\overline{w}) + 2L(\overline{w})L(z) - 2L(z \circ \overline{w}) + P(z)P(\overline{w}).$$

Moreover

$$\tilde{B}(z, w) = P(z)P(z^{-1} - \overline{w}) = P(\overline{w}^{-1} - z)P(\overline{w}),$$

provided that z^{-1}, resp. w^{-1}, exists.

Proof. (2) yields

$$\tilde{p}(z) - \overline{\tilde{p}(w)} = 2i \left(-c + (c - z)^{-1} + (c - \overline{w})^{-1} \right) = 2i \left(j(c - z^{-1}) - j(c - \overline{w}) \right)$$

provided that z^{-1} exists. The definition of $\tilde{B}$ now shows

$$\begin{aligned}
\tilde{B}(z, w) &= -\frac{1}{4} P(c - z) P \left(\tilde{p}(z) - \overline{\tilde{p}(w)} \right) P(c - \overline{w}) \\
&= -P(c - z) B(j(u), j(v)) P(\overline{v}) \\
&= -P(c - z) \frac{\partial j(u)}{\partial u} B_j(u, v) \overline{\frac{\partial j(v)}{\partial v}} P(\overline{v}),
\end{aligned}$$

with $u = c - z^{-1}$, $v = c - w$. Lemma 3 implies

$$\tilde{B}(z, w) = -P(c - z)P^{-1}(c - z^{-1})B(u, v).$$

In view of the formulas $B(u, v) = -P(z^{-1} - \overline{w})$ as well as $P(c - z^{-1}) = P^{-1}(z)P(c - z)$ proved in §1, (7), we get for $M_1 z = p(z)$ and $M_2 z = j(z)$

$$\tilde{B}(z, w) = P(z)P(z^{-1} - \overline{w}).$$

Equation (7) in §1 now states

$$\tilde{B}(z, w) = P(\overline{w}^{-1} - z)P(\overline{w}).$$

Considering the mutation $\mathcal{A}(i)_{\overline{w}}$, we know that $\overline{w}^{-1}$ is the unit element of this algebra, and that $P_{\overline{w}}(u) = P(u)P(\overline{w})$ is its quadratic representation. Hence

$$\tilde{B}(z, w) = P_{\overline{w}}(\overline{w}^{-1} - z) = \mathrm{Id} - 2L_{\overline{w}}(z) + P_{\overline{w}}(z).$$

The explicit formula for $L_{\overline{w}}(z)$ (see Chapter IV, §2) yields the remaining formula for $\tilde{B}(z, w)$. $\square$

Lemma 5. *If the map $z \mapsto g(z)$ belongs to the group $\tilde{H}_0$, then $z \mapsto g(z)$ is a linear transformation.*

Proof. By definition the map $z \mapsto g(z)$ belongs to $\tilde{H}_0$ if and only if $g(0) = 0$ and

$$\tilde{B}(g(z), g(w)) = \frac{\partial g(z)}{\partial z} \tilde{B}(z, w) \overline{\frac{\partial g(w)}{\partial w}}^*.$$

It follows from Lemma 4 that $\tilde{B}(z, 0) = \text{Id}$. Therefore setting $w = 0$ shows

$$\text{Id} = \frac{\partial g(z)}{\partial z} \overline{A}^*, \quad \text{with} \quad A = \frac{\partial g(w)}{\partial w}\Big|_{w=0}.$$

Hence $\frac{\partial g(z)}{\partial z}$ is constant, thus $g(z) = Az + b$. Now $g(0) = 0$ implies $b = 0$. $\square$

We also write $A \in \tilde{H}_0$ if $z \mapsto Az$ belongs to $\tilde{H}_0$. A linear transformation A of $X(i)$ belongs to $\tilde{H}_0$ if and only if

$$\tilde{B}(Az, Aw) = A\tilde{B}(z, w)\overline{A}^*.$$

Comparing the terms which are homogeneous of degrees $0, 1, 2$ in z, we see that $A \in \tilde{H}_0$ is equivalent to

$$(4) \qquad\qquad A\overline{A}^* = \text{Id},$$

$$(5) \qquad\qquad L_{\overline{Aw}}(Az) = AL_{\overline{w}}(z)\overline{A}^*,$$

with $L_u(z) = L(z)L(u) - L(u)L(z) + L(u \circ z)$, and

$$P(Az)P(\overline{Aw}) = AP(z)P(\overline{w})\overline{A}^*.$$

In this identity let $w = c$, then $A \in \Gamma(\mathcal{A}(i))$ follows, hence $P(Az) = AP(z)A^*$ by Theorem IV.5. Since $\overline{A}$ belongs to $\Gamma(\mathcal{A}(i))$ whenever A does (as is shown by a simple computation), a linear transformation A belongs to $\tilde{H}_0$ if and only if (4), (5) hold and if in addition

$$(6) \qquad\qquad P(Az) = AP(z)A^*$$

is satisfied.

It is easy to see that the map $z \mapsto Vz$ belongs to $\tilde{H}_0$ if V is an automorphism of the Jordan algebra $\mathcal{A}$, since in this case we have $L(Vz)V = VL(z)$ and $VV^* = \text{Id}$. Conversely, we find

Lemma 6. *Let $A \in \tilde{H}_0$ and $Ac = c$. Then A is an automorphism of the Jordan algebra $\mathcal{A}$. In particular, A is real.*

Proof. It follows from (4) that $Ac = \overline{A}c = A^*c = c$ holds. Applying (5) to c and using $L_u(v)c = u \circ v$, it follows that $Az \circ \overline{Aw} = A(z \circ \overline{w})$. An application of (6) shows $Az \circ Az = Az^2$. Here we substitute $z + \overline{w}$ for z and compare the linear terms: $Az \circ A\overline{w} = A(z \circ \overline{w})$. Together, these formulas show that $\overline{A} = A$, and that A is an automorphism of $\mathcal{A}$.   □

In order to give a description of $\tilde{H}_0$, we use the "unitary" elements of $\mathcal{A}(i)$, i.e. the subset Q of $\mathcal{A}(i)$ containing those elements q for which the inverse exists and is equal to $\overline{q}$. An element $q \in X(i)$ thus belongs to Q if and only if

$$q \circ \overline{q} = c, \quad \text{and} \quad q, \overline{q} \quad \text{commute.}$$

Since $\overline{q}$ is the inverse of q, we conclude

(7) $$P(q)P(\overline{q}) = \text{Id} \quad \text{for} \quad q \in Q.$$

On the other hand, q and $\overline{q}$ commute if and only if a and b commute, where $q = a + ib$. Hence there is a bijective correspondence between Q and the set of solutions $(a, b) \in \mathcal{A} \times \mathcal{A}$ of

$$a^2 + b^2 = c, \quad a, b \quad \text{commute.}$$

Theorem 1. *A linear transformation A belongs to $\tilde{H}_0$ if and only if there exists an element $q \in Q$ such that $A = P(q)V$, with V an automorphism of the Jordan algebra $\mathcal{A}$. Furthermore, Ac belongs to Q if $A \in \tilde{H}_0$.*

Proof. We divide the proof into several parts.

(i) $P(q)V$ *belongs to* $\tilde{H}_0$.

Since every automorphism V belongs to $\tilde{H}_0$, we only need to prove that $P(q)$, $q \in Q$, belongs to $\tilde{H}_0$, i.e. that $\tilde{B}(P(q)z, P(q)w) = P(q)\tilde{B}(z, w)\overline{P(q)}$ holds. Both sides are polynomials in z. Hence we may assume that z^{-1} exists. Using Lemma 4, the fundamental formula, and (7) we conclude $(P(q)z)^{-1} = P(\overline{q})z^{-1}$ and

$$\tilde{B}(P(q)z, P(q)w) = P(P(q)z)P\left(P(\overline{q})z^{-1} - P(\overline{q})\overline{w}\right)$$
$$= P(q)P(z)P(z^{-1} - \overline{w})P(q) = P(q)\tilde{B}(z, w)P(\overline{q}).$$

(ii) $a = Ac \in Q$ *if* $A \in \tilde{H}_0$.

It follows from (6) that the inverse of a exists. Substituting $z = w = c$ in (5) we get $L_{\overline{a}}(a) = A\overline{A}^* = \text{Id}$ or $\overline{a} = a^{-1}$.

(iii) *If* $a \in Q$ *then there is an element* $q \in Q$ *such that* $a = q^2$.

Let

$$a = \left(\sum_\nu \lambda_\nu c_\nu\right) \circ (c + v),$$

be the minimal decomposition. In view of $\bar{a} = a^{-1}$ we obtain

$$\left(\sum_\nu \bar{\lambda}_\nu \bar{c}_\nu\right) \circ (c + \bar{v}) = \left(\sum_\nu \frac{1}{\lambda_\nu} c_\nu\right) \circ (c + v)^{-1}.$$

The uniqueness of the minimal decomposition shows that there is a permutation ρ such that

$$\bar{c}_\nu = c_{\rho(\nu)}, \quad \bar{\lambda}_\nu = \frac{1}{\lambda_{\rho(\nu)}}, \quad c + \bar{v} = (c + v)^{-1}.$$

Choose complex numbers η_ν such that

$$\eta_\nu^2 = \lambda_\nu \quad \text{and} \quad \bar{\eta}_\nu = \frac{1}{\eta_{\rho(\nu)}}.$$

Since $c_\nu, \bar{c}_\nu, v, \bar{v}$ belong to $C[a]$, the element

$$q = \left(\sum_\nu \eta_\nu c_\nu\right) \circ \left(\sum_{m=0}^\infty \binom{1/2}{m} v^m\right)$$

belongs to Q and satisfies $a = q^2$.

(iv) If $A \in \tilde{H}_0$ then $A = P(q)V$, with $q \in Q$ and V an automorphism of A.

It follows from (ii) that $a = Ac \in Q$. Find $q \in Q$ such that $a = q^2$ (see (iii)) and consider the element $V = P^{-1}(q)A$. We get $Vc = P^{-1}(q)q^2 = c$. Thus V is an automorphism of A by Lemma 6. $\qquad\square$

Corollary 1. *Let $A \in \tilde{H}_0$, then $\bar{A}$, A^* and $e^{i\gamma}A$, γ real, belong to $\tilde{H}_0$.*

§3. Application to the set H

Starting with an element f in H_{ic} the map $g = p \cdot f \cdot \tilde{p}$ belongs to the isotropy group $\tilde{H}_0$ and Lemma 5 states that $g(z) = Az$ is a linear transformation. In a neighborhood of ic we therefore obtain

(1) $\qquad\qquad f \in H_{ic} \Leftrightarrow f(z) = \tilde{p}(Ap(z)), \quad A \in \tilde{H}_0.$

This implies

(2) $\qquad\qquad \left.\dfrac{\partial f(z)}{\partial z}\right|_{z=ic} = A,$

and moreover every function of the form (1) is a rational function. In view of $H = H^\infty \cdot H_{ic} \cdot H^\infty$ we find

Theorem 2. *Every $f \in H$ can be extended to a rational function on the entire vector space $X(i)$. Each $f \in H_{ic}$ is uniquely determined by the value of $\frac{\partial f(z)}{\partial z}$ in any given point of its domain of definition.*

Sometimes it is important to know under which conditions the Jacobian of a map is a constant.

Theorem 3. *Let $f \in H$. Then the Jacobian $\det \frac{\partial f(z)}{\partial z}$ is constant if and only if f belongs to H^{∞}.*

Proof. Obviously, the Jacobian is constant for $f \in H^{\infty}$. Conversely, let the Jacobian of $z \mapsto f(z)$ be a constant. Without restriction we may assume that $f \in H_{ic}$. Formula (1) shows that

$$\det \frac{\partial \tilde{p}(w)}{\partial w}\bigg|_{w=Ap(z)} \cdot \det \frac{\partial p(z)}{\partial z} = \alpha$$

is independent of z. Since $\tilde{p}$ is the inverse map of p we get

$$\det \frac{\partial \tilde{p}(w)}{\partial w}\bigg|_{w=Ap(z)} = \alpha \cdot \det \frac{\partial \tilde{p}(w)}{\partial w}\bigg|_{w=p(z)}.$$

Since both sides are rational functions, and p is locally invertible, we obtain

$$\det \frac{\partial \tilde{p}(w)}{\partial w}\bigg|_{w=Au} = \alpha \cdot \det \frac{\partial \tilde{p}(w)}{\partial w}\bigg|_{w=u}$$

for all u in an open dense subset of $X(i)$. Using equation (2) in §2 yields

$$\det P(c - u) = \alpha \det P(c - Au)$$

(identically in u). Since $\det P(c - u) = 1 - 2\tau(c, u) + \ldots$, it follows that $\alpha = 1$, and $\tau(c, u) = \tau(c, Au)$, hence $c = A^{*}c$. By Corollary 1, A^{*} belongs to $\tilde{H}_{0}$ whenever A does. Thus we conclude from Lemma 6 that A is an automorphism of $\mathcal{A}$. Hence $f(z) = \tilde{p}(Ap(z)) = A\tilde{p}(p(z)) = Az$ follows. $\square$

Given $a \in X$, we consider the translation

$$t_{a}(z) = z + a,$$

and obtain $t_{a} \in H^{\infty}$ for every $a \in X$. Denote by T the subgroup of H^{∞} containing these translations. Lemma 2(a) yields

$$H^{\infty} = \Sigma \cdot T = T \cdot \Sigma.$$

Moreover

(3) $$j \cdot H^{\infty} \subset H^{\infty} \cdot j \cdot T.$$

To verify this, it is sufficient to prove $j\Sigma \subset \Sigma j$. But this follows from $(Wz)^{-1} = W^{*-1}z^{-1}$, whenever $P(Wz) = WP(z)W^*$.

In the case $X = \mathbf{R}$ we have the relation $(jt_c)^3 = \mathrm{Id}$. As an analog we obtain this relation in the general case (since $\mathbf{C}[a]$ is commutative and associative), and moreover

Lemma 7. *A relation*

$$t_a \cdot j \cdot t_b \cdot j \cdot t_{a'} \cdot j = W, \quad \text{with} \quad a, a', b \in X, \quad W \quad \text{linear transformation,}$$

holds if and only if the inverse of b exists and $a = a' = b^{-1}$, $W = P(a)$.

Proof. In case $a' = a$, $b = a^{-1}$, $W = P(a)$ we get the relation immediately from the identity

$$(4) \qquad (u^{-1} - v^{-1})^{-1} = u - P(u)(u - v)^{-1},$$

which has been proved in Theorem V.4. Conversely, the relation in question is equivalent to

$$(\dagger) \qquad -(b - z^{-1})^{-1} = -a - W(z - a')^{-1}.$$

We compute $\frac{\partial}{\partial z}$ on both sides of this identity. Thus we obtain

$$P(z)P(b - z^{-1})W = P(z - a').$$

Here we substitute λz for z, then $\lambda \to \infty$ gives $P(b)W = \mathrm{Id}$. Hence the inverse of b exists, and identity (7) in §1 yields $P(b^{-1} - z) = P(a' - z)$. We substitute $z = 0$, and after that $z = c$. Thus we get $L(a') = L(b^{-1})$, hence $a' = b^{-1}$. Substitute this, and $W = P(b)^{-1}$, into $(\dagger)$ and put $z = 2b^{-1}$ to find $a = b^{-1} = a'$.   □

The relation $(t_c j)^3 = \mathrm{Id}$ in particular yields

$$(5) \qquad j \in T \cdot j \cdot T \cdot j \cdot T.$$

After these preliminaries we are able to prove a structure theorem for the set H.

Theorem 4. $H = H^\infty \cdot j \cdot T \cdot j \cdot T = \overline{H^\infty \cdot j \cdot T}$,

where the bar denotes the closure in H with respect to the topology of compact uniform convergence in H.

Proof. We divide the proof into several steps and denote by N the subset of H_{ic} containing those maps f which satisfy

$$\det P(c - Ac) \neq 0, \quad \text{with} \quad A = \left.\frac{\partial f(z)}{\partial z}\right|_{z=ic}.$$

(i) $H_{ic} \subset N \cdot (H^\infty \cup H^\infty \cdot j \cdot H^\infty)$.

Let $f \in H_{ic}$. Then we get a representation $f(z) = \tilde{p}(Ap(z))$, with $A \in \tilde{H}_0$. Now let

$$M = \begin{pmatrix} \cos\varphi & \sin\varphi \\ -\sin\varphi & \cos\varphi \end{pmatrix} \in GL(2, \mathbf{R}).$$

Then Lemma 3 states that the generalized Möbius transformation $z \mapsto Mz$ belongs to H, and it maps ic to ic, hence $M \in H_{ic}$. It follows, using (1) and (2), that $Mz = \tilde{p}(e^{2i\varphi}p(z))$ and hence

$$f(Mz) = \tilde{p}(e^{2i\varphi} Ap(z)).$$

In view of $\det P(c - e^{2i\varphi} Ac) = 0$ only for a finite number of real $\varphi \in [0, 2\pi]$ there exists a φ such that $f(Mz) = g(z)$ belongs to N. Equation (1) in §1 states that $z \mapsto Mz$ belongs to $H^\infty \cup H^\infty \cdot j \cdot H^\infty$, and this proves the above inclusion.

(ii) $H_{ic} = \overline{N}$.

In the notation of (i) we find a sequence φ_m converging to 0 such that the maps

$$z \mapsto f_m(z) = \tilde{p}(e^{2\pi i\varphi_m} Ap(z))$$

belong to N. Obviously f_m converges uniformly to f on compact subsets contained in a suitable neighborhood of ic.

(iii) $N \subset H^\infty \cdot j \cdot H^\infty$.

Let $f(z) = \tilde{p}(Ap(z))$ be an element in N, and $r = A^{-1}c$. From $A \in \Gamma(\mathcal{A}(i))$ we get $\det P(c - r) \neq 0$. We know from Theorem 1 that $r \in Q$, i.e. $\bar{r} = r^{-1}$. Hence $\tilde{r} = \tilde{p}(r)$ is defined, and equation (2) in §2 states that

$$\bar{\tilde{r}} - \tilde{r} = 2ic - 2i(c - \bar{r})^{-1} - 2i(c - r)^{-1} = 0,$$

hence $\tilde{r}$ is real. Using equations (1) and (2) in §2 as well as (4) we obtain

$$\begin{aligned} f(z) &= -ic + 2i(c - Ap(z))^{-1} = -ic + 2i\overline{A}(r - p(z))^{-1} \\ &= -ic - \overline{A}\left(\tilde{r} + ic - P(\tilde{r} + ic)(\tilde{r} - z)^{-1}\right) = Wj(z - \tilde{r}) + t, \end{aligned}$$

with $W = \overline{A}P(\tilde{r} + ic)$ and $t = -ic - \overline{A}(\tilde{r} + ic)$. Since $\tilde{r}$ is real and the map $z \mapsto j(z)$ belongs to H (see Lemma 3), it follows that $z \mapsto Wz + t$ belongs to

H. Lemma 2(a) states that this map belongs to H^∞, i.e. $W \in \Sigma$ and $t \in X$. (This can also be proved directly using the definition of W and t.) Hence $f \in H^\infty \cdot j \cdot H^\infty$ follows.

(iv) $j \cdot T \cdot j \subset H$.

Since the elements on the left hand side leave the linear transformation B invariant, by Lemma 1 we only need to prove that $f = j \cdot t_a \cdot j$ maps some open subset of Z into Z. We obtain

$$f(z) = -(a - z^{-1})^{-1}.$$

Let $z = x + ic$ and $x \in \mathcal{A}$. Then z belongs to Z, and it is sufficient to prove that there is an $x \in \mathbf{R}[a]$ satisfying $f(x + ic) \in Z$. Using the associativity of $\mathbf{C}[a]$, we get

$$\mathrm{Im}(f(z)) = ((a \circ x - c)^2 + a^2)^{-1}$$

and we need to show that there is an $x \in \mathbf{R}[a]$ that satisfies $(a \circ x - c)^2 + a^2 \in Y$. If the inverse of a exists we take $x = a^{-1}$, and obtain $a^2 \in Y$ from Theorem VI.4. Otherwise we consider the minimal decomposition of a in $\mathcal{A}(i)$:

$$a = \sum_k \lambda_k c_k + v, \quad v \text{ nilpotent}, \quad c_1 + \ldots + c_r = c.$$

Since a is real, $v \in \mathbf{R}[a]$ follows, and the λ_k and c_k are real resp. pairwise conjugate. Let $\lambda_1 = 0$, then the other eigenvalues are not zero. Put

$$x = \sum_{k>1} \frac{1}{\lambda_k} c_k,$$

then x is real and $a \circ x = c_2 + \ldots + c_r + v'$, v' nilpotent, and $(a \circ x - c)^2 = c_1 + v''$, v'' nilpotent. For

$$b = c_1 + \sum_{k>1} \lambda_k c_k$$

it follows that $(a \circ x - c)^2 + a^2 = b^2 + v'''$, and b and the nilpotent v''' are real. Moreover the inverse of b exists. Hence $b^2 \in Y$, and in view of $\det P(b^2 + \lambda v''') \neq 0$ for $0 \leq \lambda \leq 1$ we get $b^2 + v''' \in Y$.

(v) *Proof of the Theorem.*

It follows from Lemma 2(c) that $H = H^\infty \cdot H_{ic} \cdot H^\infty$, moreover (i) and (iii) state

$$H_{ic} \subset H^\infty \cdot j \cdot H^\infty \cup H^\infty \cdot j \cdot H^\infty \cdot j \cdot H^\infty.$$

Equation (5) shows that the first set on the right hand side is contained in the second one. Finally (3) states $H_{ic} \subset H^\infty \cdot j \cdot T \cdot j \cdot T$ and therefore $H \subset H^\infty \cdot j \cdot T \cdot j \cdot T$. The other inclusion follows from (iv).

For the second part we use (ii) and (iii)

$$H = H^\infty \cdot H_{ic} \cdot H^\infty = H^\infty \cdot \overline{N} \cdot H^\infty \subset \overline{H^\infty \cdot j \cdot H^\infty} \subset \overline{H} = H.$$

$\square$

§4. Biholomorphic automorphisms of half-spaces

A biholomorphic map $f : Z \to Z$ is called a *biholomorphic automorphism* of Z. We now consider the subset $H(Z)$ consisting of all biholomorphic automorphisms $f \in H$. Thus a biholomorphic map $f : Z \to Z$ belongs to $H(Z)$ if and only if

$$(1) \qquad \beta(f(z)) = \left| \det \frac{\partial f(z)}{\partial z} \right|^2 \beta(z), \quad \text{where} \quad \beta(z) = \det P(z - \overline{z}).$$

Obviously $H(Z)$ is a group, and Lemma 2 shows that H^∞ is a subgroup of $H(Z)$.

The restrictive condition (1) is quite natural. In fact, $H(Z)$ can be characterized as follows:

$H(Z)$ *is the (uniquely determined) maximal subgroup of the group of all biholomorphic automorphisms of Z containing the maps*

$$z \mapsto Wz, \quad z \mapsto z + a \quad \text{and} \quad W \in \Sigma, \quad a \in X,$$

and admitting an invariant volume element.

Indeed, let H' be such a group with respect to the invariant volume element $dv(z) = \gamma(z)dx\,dy$. Then $\gamma(z)$ is a complex-valued function and $dv(z)$ is invariant if and only if $\gamma(f(z))d(\mathrm{Re}(f(z)))d(\mathrm{Im}(f(z))) = \gamma(z)dx\,dy$. This is equivalent to

$$\gamma(f(z)) \cdot \left| \det \frac{\partial f(z)}{\partial z} \right|^2 = \gamma(z).$$

Forming $\alpha(z) := \beta(z) \cdot \gamma(z)$ we see that $\alpha(z)$ is invariant under H', and in particular under H^∞. Since H^∞ acts transitively on Z we conclude that $\alpha(z)$ is constant, and hence $H' \subset H(Z)$. Thus $dv(z) = \alpha \det P^{-1}(z - \overline{z})dx\,dy$ is the invariant volume element.

The theory of Bergman kernels (see Appendix, Theorem 4) immediately yields

Theorem 5. *If there is at least one square integrable holomorphic function $f \neq 0$ on Z, then $H(Z)$ coincides with the entire group of all biholomorphic automorphisms of Z.*

In general the map $z \mapsto j(z) = -z^{-1}$ does not belong to $H(Z)$. However, Theorem 4 implies

Theorem 6. *The group $H(Z)$ of all biholomorphic automorphisms of Z satisfying (1) is a subset of $\overline{H^\infty \cdot j \cdot T}$. Let $\hat{T}$ be the subgroup of T containing the translations t_a such that $j \cdot t_a \cdot j$ is a biholomorphic automorphism of Z. Then*

$$H(Z) = H^\infty \cdot j \cdot \hat{T} \cdot j \cdot T.$$

Corollary 2. *If the map $z \mapsto j(z)$ is a biholomorphic automorphism of Z, then j belongs to $H(Z)$, and*

$$H(Z) = H^\infty \cdot j \cdot T \cdot j \cdot T = \overline{H^\infty \cdot j \cdot T}.$$

In particular $H(Z)$ is generated by the maps

$$z \mapsto -z^{-1}, \quad z \mapsto z + a \quad (a \in X), \quad \text{and} \quad z \mapsto Wz \quad (W \in \Sigma).$$

Proof. We only need to prove the last statement of Theorem 6. Since H^∞ and T are subgroups of $H(Z)$, we obtain $H(Z) \subset H^\infty \cdot j \cdot \hat{T} \cdot j \cdot T \subset H(Z)$ from Theorem 4. $\qquad\square$

Theorem 7. *Assume that $z \mapsto j(z)$ belongs to $H(Z)$. Then*

(a) $H(Z)$ is generated by $z \mapsto j(z)$, $z \mapsto z + a$, and $z \mapsto Vz$, where $a \in X$ and V is an automorphism of the Jordan algebra $\mathcal{A}$.

(b) $H(Z)$ does not contain more connected components than $\text{Aut}(\mathcal{A})$.

(c) If each automorphism of $\mathcal{A}$ is a product of transformations $P(u)$, $u \in X$ with $\det P(u) \neq 0$, then $H(Z)$ is generated by $z \mapsto j(z)$ and $z \mapsto z + a$, where $a \in X$. Thus $H(Z)$ is connected.

Proof. (a) Theorem 6 states that we only need to prove that each map of the form $z \mapsto Wz$, $W \in \Sigma$, is a product of the mappings indicated above. From Theorem VI.5 we know that $W = P(a)V$ for some $a \in \mathcal{A}$ and some automorphism V of the Jordan algebra $\mathcal{A}$. It follows from Lemma 7 that $P(a)$ is a product of the maps $z \mapsto j(z)$ and $z \mapsto z + a$ as well as $z \mapsto z + a^{-1}$.

(b) Part (a) shows that any element $f \in H(Z)$ can be written as $f = Vg$, where V is an automorphism of $\mathcal{A}$ and g belongs to the group $\tilde{H}$ generated by j and T. Now the curve $\xi \mapsto t_{\xi a}$, $0 \leq \xi \leq 1$, joins Id and t_a in $\tilde{H}$. We conclude that the mappings $z \mapsto Mz$, $M \in SL(2, \mathbf{R})$, belong to $\tilde{H}$. Thus $\xi \mapsto M_\xi = \begin{pmatrix} \cos \pi\xi & \sin \pi\xi \\ -\sin \pi\xi & \cos \pi\xi \end{pmatrix}$, $0 \leq \xi \leq 1$, joins Id and j in $\tilde{H}$. Therefore $\tilde{H}$ is connected. Let A_k denote the components of the group of automorphisms of

$\mathcal{A}$, then $A_k \tilde{H}$ is connected, and $H(Z)$ is the union of the sets $A_k \tilde{H}$. Hence the components of $H(Z)$ are the different sets among the $A_k \tilde{H}$.

(c) Let V be an element of $Aut(\mathcal{A})$. Then V is a product of some $P(a)$. We apply Lemma 7 again, and we conclude that V is a product of elements as indicated in the assertion. □

Without proof (cf. U. Hirzebruch [5], Satz 10, for formally real Jordan algebras) we mention

Theorem 8. *Let $\mathcal{A} = \mathcal{A}_1 \oplus \ldots \oplus \mathcal{A}_r$ be the decomposition of $\mathcal{A}$ into simple Jordan algebras and let $Z_1, \ldots, Z_r$ be the half-spaces of $\mathcal{A}_1, \ldots, \mathcal{A}_r$. Then Z is the direct product of $Z_1, \ldots, Z_r$, and a map $f \in H(Z)$ can be written as $f_0 \cdot f_1$, where f_0 is a permutation of $Z_1, \ldots, Z_r$ and f_1 belongs to the direct product of the groups $H(Z_1), \ldots, H(Z_r)$.*

Let us now show that the requirement $j \in H(Z)$ yields a strong restriction on the Jordan algebra $\mathcal{A}$:

Lemma 8. *If the map $z \mapsto j(z) = -z^{-1}$ is defined on Z, then $\mathcal{A}$ is a formally real Jordan algebra.*

Proof. The hypothesis states that z^{-1} exists for $z \in Z$, i.e. that $\det P(z) \neq 0$ for $z \in Z$. Since there is an element $W \in \Sigma \subset \Gamma(\mathcal{A})$ such that $Wz = u + ic$, $u \in X$, we conclude that the inverse of $u + ic$ exists for every $u \in X$. We will show that then necessarily the eigenvalues of every element $u \in X$ are real. If at least one eigenvalue of some u is not real, then the minimal decomposition of u (see Chapter V, §§3 and 6) yields an idempotent e in $\mathcal{A}(i)$ such that $\bar{e} \neq e$, $e \circ \bar{e} = 0$. The element $x = i(e - \bar{e})$ is real and $x + ic$ does not possess an inverse. (Note that e and $\bar{e}$ are contained in an associative subalgebra, and that $(x + ic) \circ (e + \bar{e}) = 2ie$ is not invertible.) This shows that the eigenvalues of $u \in X$ are real. Hence the eigenvalues of $L(u^2)$ are non-negative (see Chapter V, §5), and we get $\tau(u, u) = \operatorname{tr} L(u^2) \geq 0$. Since τ is non-singular, it follows that τ is positive definite, and Theorem VI.12 states that $\mathcal{A}$ is formally real. □

In the next section we will see that for any formally real Jordan algebra the map $j : z \mapsto -z^{-1}$ belongs to $H(Z)$.

§5. Formally real Jordan algebras

Let us first consider the involution $j(z)$ for arbitrary semisimple Jordan algebras. Since Σ acts transitively on Y, for every $z \in X(i)$ there are $W \in \Sigma$ and $u \in X$ such that

$$z = W(u + ic).$$

Hence $\det P(z) \neq 0$ if and only if $\det P(u^2 + c) = \det(P(u+ic)P(u-ic)) \neq 0$, and

(1) $\quad j(z) = -W^{*-1}(u + ic)^{-1} = W^{*-1}(-(u^2 + c)^{-1} \circ u + i(u^2 + c)^{-1}),$

because the subalgebra $C[u]$ is associative. In particular, using the fundamental formula we get for $x = Wu$, $y = Wc$:

(2) $\quad \mathrm{Im}\,(j(z)) = (y + P(x)y^{-1})^{-1}, \quad \text{if} \quad z \in Z, \quad \det P(z) \neq 0.$

Lemma 8 states that $\mathcal{A}$ is formally real if $j(z)$ is defined on Z. On the other hand, let $\mathcal{A}$ be formally real. Then Y is a domain of positivity (see Chapter VI, §4), and therefore convex. Given $y \in Y$ we have $y^{-1} \in Y$, $P(x)y^{-1} \in \overline{Y}$, and $y + P(x)y^{-1} \in Y$. Hence $\det P(z) \neq 0$ and $\mathrm{Im}\,(j(z)) \in Y$ follow. The map $z \mapsto j(z)$ therefore maps Z into Z. In view of $j^2 = \mathrm{Id}$, this map is a biholomorphic automorphism. From Lemma 3 it follows that $j \in H(Z)$. Summarizing we obtain:

Theorem 9. *Given a semisimple Jordan algebra $\mathcal{A}$, the map $z \mapsto j(z) = -z^{-1}$ belongs to $H(Z)$ if and only if $\mathcal{A}$ is formally real.*

For the rest of this chapter *we assume that the given Jordan algebra $\mathcal{A}$ is formally real*. From Chapter VI, §4, we know that Y is then a domain of positivity, and $\Sigma = \Sigma(Y)$ coincides with the entire group of linear automorphisms of Y.

Now we again (see §2) consider the mappings

(3) $\qquad p(z) = (z - ic) \circ (z + ic)^{-1} = c - 2i(z + ic)^{-1},$

(4) $\qquad \tilde{p}(z) = i(c + z) \circ (c - z)^{-1} = -ic + 2i(c - z)^{-1}.$

Given z in the closure $\overline{Z}$ of Z, we get $z + ic \in Z$, and Theorem 9 yields $\det P(z + ic) \neq 0$. Hence $p(z)$ is holomorphic on $\overline{Z}$. Denote by $\tilde{Z}$ the image of Z under p:

$$\tilde{Z} = \{p(z); z \in Z\}.$$

Then $\tilde{p}$ maps $\tilde{Z}$ onto Z and both maps are biholomorphic and inverse to each other.

Theorem 10. *The domain $\tilde{Z}$ is bounded, and the mappings $z \mapsto \bar{z}$ as well as $z \mapsto Az$, $A \in \tilde{H}_0$, map $\tilde{Z}$ onto itself.*

Proof. Since Y is a domain of positivity we have the orderings ">" and "≥" on X, cf. Chapter I, §2. From Chapter VI, §6, we know that for a function $f(\tau)$ of a real variable τ which maps an interval $]\alpha; \beta[$ into $]\gamma; \delta[$, the relations $\gamma c < f(u) < \delta c$ hold for $\alpha c < u < \beta c$. An analogous statement is true if we replace "<" by "≤". Now let $f(\tau) = -\tau(\tau^2 + 1)^{-1}$ and $g(\tau) = (\tau^2 + 1)^{-1}$. Since f resp. g map the real line onto the closed interval $\left[-\frac{1}{2}; \frac{1}{2}\right]$ resp. onto the interval $]0; 1]$, we conclude that

$$-\frac{1}{2}c \le -(u^2 + c)^{-1} \circ u \le \frac{1}{2}c, \quad 0 < (u^2 + c)^{-1} \le c$$

for every $u \in X$. Since $u \le v$ and $W \in \Sigma$ implies $Wu \le Wv$, we get from (1)

(5) $\quad -\frac{1}{2}y^{-1} \le \operatorname{Re}(j(z)) \le \frac{1}{2}y^{-1}, \quad 0 < \operatorname{Im}(j(z)) \le y^{-1} \quad$ for $\ z \in Z,$

using $W^{*-1}c = (Wc)^{-1} = y^{-1}$ in the notation of (1). By virtue of a similar argument we conclude $0 < (c + y)^{-1} < c$, whenever $iy \in Z$. It follows from (3) that $p(z) = c + 2i \cdot j(z + ic)$, and hence for $z \in Z$

(6) $\qquad\qquad -c < \operatorname{Re}(p(z)) < c, \quad -c < \operatorname{Im}(p(z)) < c.$

From equation (9) in Chapter I, §2 we know that in a domain of positivity the set $\{u \in Y; 0 < a \le u \le b\}$ is a compact. Hence $\tilde{Z}$ is bounded.

Obviously, the map $z \mapsto -\bar{z}$ is a bijection of Z. Given $w = p(z) \in \tilde{Z}$ we obtain

$$\overline{w} = \overline{p(z)} = c + 2i(\bar{z} - ic)^{-1} = c - 2i(-\bar{z} + ic)^{-1} = p(-\bar{z}) \in \tilde{Z},$$

i.e. $w \in \tilde{Z}$ implies $\overline{w} \in \tilde{Z}$.

The remaining statement follows from the fact that $z \mapsto \tilde{p}(Ap(z))$, $A \in \tilde{H}_0$, maps Z onto itself. $\qquad\square$

Combining the results of the Theorems 5, 6, 7, 9 and 10 we obtain the following main result about the group of biholomorphic automorphisms of Z:

Theorem 11. *Let A be a formally real Jordan algebra and let Z be the half-space associated with A. Then:*

(a) $H(Z) = H^\infty \cdot j \cdot T \cdot j \cdot T = \overline{H^\infty \cdot j \cdot T}$ *is equal to the entire group of all biholomorphic automorphisms of Z.*

(b) $H(Z)$ *is generated by the mappings $z \mapsto j(z) = -z^{-1}$, $z \mapsto z + a$, $z \mapsto Vz$, with $a \in X$ and V an automorphism of the Jordan algebra A.*

(c) $H(Z)$ *does not contain more connected components than $\operatorname{Aut}(A)$.*

(d) *If every automorphism of A can be represented as a product of some transformations $P(u)$, $u \in X$ invertible, then $H(Z)$ is generated by the maps $z \mapsto z + a$, $a \in X$, and $z \mapsto -z^{-1}$. In this case $H(Z)$ is connected.*

§6. The bounded symmetric domain $\tilde{Z}$

Again let $\mathcal{A}$ be a formally real Jordan algebra, Z the half-space associated with $\mathcal{A}$ and $\tilde{Z}$ the image of Z under $z \mapsto p(z)$. In order to give a description of the bounded domain $\tilde{Z}$ we need a new description of the unitary elements in $X(i)$ (see §2):

Lemma 9. $Q = \{e(ix); x \in X\} = e(iX)$, and Q is a compact subset of $X(i)$.

Proof. From $e^{-1}(a) = e(-a)$ we conclude $e(ix) \in Q$. Let now $q \in Q$. Then we know from the proof of Theorem 4 that $u = \tilde{p}(q)$ is real if $(c-q)^{-1}$ exists. Hence $q = p(u)$ follows. Let $u = \Sigma_k \lambda_k c_k$ be the minimal decomposition of u, then λ_k, c_k are real. Since $p(\lambda_k) \in \mathbf{C}$ has the absolute value 1, we obtain

$$q = \sum_k p(\lambda_k)c_k = \sum_k e^{i\varphi_k} c_k = e(ix), \quad \text{with} \quad x = \sum_k \varphi_k c_k$$

for some real φ_k. Hence $q = e(ix)$ holds provided that $\det P(c-q) \neq 0$. For an arbitrary $q \in Q$ there exists a sequence ψ_ν converging to 0 such that $\det P(c - e^{i\psi_\nu} \cdot q) \neq 0$. Hence $e^{i\psi_\nu} q = e(ix_\nu)$, and the sequence x_ν can be chosen to be convergent.

In the proof of Theorem 1 we saw that for any $q \in Q$ there exists an element $p \in Q$ such that $q = p^2$. Since $P(p)P(\bar{p}) = \mathrm{Id}$, and since $P(p)$ is self-adjoint with respect to the positive definite bilinear form $\tau(x,y) = \mathrm{tr}\, L(x \circ y)$ we conclude that $\{P(p); \ p \in Q\}$ is bounded. Hence $Q = \{P(p)c; \ p \in Q\}$ is bounded, and thus Q is compact. □

This description of Q gives rise to an analog of polar coordinates.

Lemma 10. *For every $z \in X(i)$ there exist $q = e(ix) \in Q$ and $r \geq 0$ (i.e. $r \in \overline{Y}$) such that $z = P(q)r$.*

Proof. For $z \in X(i)$ let us consider the function

$$\varphi(q) = |\mathrm{Im}\,(P(q)z)| \quad \text{for} \quad q \in Q,$$

where $|a|^2 = \tau(a,a)$. Since τ is positive definite, $|a|$ is a norm on X. As $\varphi(q)$ is continuous on Q and Q is compact, we find $q_0 \in Q$ such that $\varphi(q) \geq \varphi(q_0)$ for every $q \in Q$. Let $w = P(q_0)z = u + iv$. Then we get $\varphi(q_0) = |v|$. It follows from Theorem 1 that there exist an automorphism V of $\mathcal{A}$ and $\tilde{q} \in Q$ such that

$$P(q)P(q_0) = VP(\tilde{q}).$$

In view of $|Va| = |a|$ we obtain for $q \in Q$:

$$|\text{Im}(P(q)w)| = |\text{Im}(P(q)P(q_0)z)| = |V\text{Im}\,(P(\tilde{q})z)\,| = \varphi(\tilde{q}) \ge \varphi(q_0) = |v|,$$

and thus $|\,\text{Im}(P(q)w)| \ge |\,\text{Im}\,(w)|$ for every $q \in Q$.

Using Lemma 9 we substitute $q = e(ix)$. Then

$$P(e(ix)) = e^{2iL(x)} = A(x) + iB(x)$$

immediately yields the relation $\text{Im}\,(P(q)w) = A(x)v + B(x)u$, whence

$$(1) \qquad |A(x)v + B(x)u|^2 \ge |v|^2 \quad \text{for every} \quad x \in X.$$

From $P(q)P(\tilde{q}) = \text{Id}$ and $P^* = P$ we conclude

$$A^2(x) + B^2(x) = \text{Id}, \; A(x)B(x) = B(x)A(x), \; A^*(x) = A(x), \; B^*(x) = B(x),$$

and (1) leads to

$$-\tau(B^2(x)v, v) + 2\tau(A(x)B(x)u, v) + \tau(B^2(x)u, u) \ge 0 \quad \text{for every} \quad x \in X.$$

Comparing the linear and quadratic terms in x and using

$$A(x) = \text{Id} - 2L^2(x) + \ldots, \; B(x) = L(x) + \ldots$$

we get $\tau(L(x)u, v) = 0$ and $\tau(L^2(x)u, u) \ge \tau(L^2(x)v, v)$ for every $x \in X$. In view of $\tau(L(x)u, v) = \tau(x \circ u, v) = \tau(x, u \circ v)$ we obtain $u \circ v = 0$. Putting $x = v$ in the second condition yields $0 \ge \tau(v^2, v^2)$, hence $v^2 = 0$. Since a formally real algebra does not contain any non-trivial nilpotent elements, we conclude $v = 0$.

So far we have proved that for a given $z \in X(i)$ there exists $q_0 \in Q$ such that $P(q_0)z = r_0$ is real. Let

$$r_0 = \sum_k \lambda_k c_k$$

be the minimal decomposition of r_0. Then λ_k and c_k are real and hence there are real φ_k such that $e^{2i\varphi_k}\lambda_k$ is non-negative. For $q_1 = \sum_k e^{i\varphi_k} c_k \in Q$ we conclude that the eigenvalues of $P(q_1)r_0$ are non-negative, whence $P(q_1)P(q_0)z \in \overline{Y}$. Using Theorem 1 again, we find an automorphism V of $\mathcal{A}$ and $q \in Q$ such that $P(q_1)P(q_0) = V^{-1}P(\tilde{q})$. In view of $V \in \Sigma$ we conclude $z = P(q)r$, $r = Vr_0 \in \overline{Y}$. $\qquad\square$

In one variable the domain $\tilde{Z}$ is equal to the unit disk. As a generalization we prove

Theorem 12. *Let $\mathcal{A}$ be a formally real Jordan algebra, let Z be the half-space associated with $\mathcal{A}$ and let $\tilde{Z}$ be the image of Z under $z \mapsto p(z)$. Then we have the following characterizations of $\tilde{Z}$:*

(a) z belongs to $\tilde{Z}$ if and only if $z = P(e(ix))r$, where $x, r \in X$, $-c < r < c$.

(b) $z \in X(i)$ belongs to $\tilde{Z}$ if and only if the hermitian linear transformation $\text{Id} - \overline{P(z)}P(z)$ is positive definite.

Proof. (a) Lemma 10 states $z = P(q)r$ for some $q \in Q$, $r \in X$. We know from Theorem 1 that $P(q)$ belongs to $\tilde{H}_0$, and hence Theorem 10 shows that $z \in \tilde{Z}$ holds if and only if $r \in \tilde{Z}$. In Chapter VI, §6, equation (2), we proved for $u \in X$:

$$(c - u) \circ (c + u)^{-1} \in Y \quad \Leftrightarrow \quad -c < u < c.$$

From the definition of $\tilde{p}$ we therefore conclude that $r \in X$ belongs to $\tilde{Z}$ if and only if $-c < r < c$.

(b) We again use $z = P(q)r$, $q \in Q$, $r \in X$, $\overline{P(q)}P(q) = \mathrm{Id}$ and the fundamental formula. This leads to $\mathrm{Id} - \overline{P(z)}P(z) = \mathrm{Id} - \overline{P(q)}P(r^2)P(q)$ and hence

(2) $$\mathrm{Id} - \overline{P(z)}P(z) > 0 \quad \Leftrightarrow \quad \mathrm{Id} - P(r^2) > 0.$$

The eigenvalues of r are real. Hence Theorem V.5 shows that the first statement in (2) is equivalent to the eigenvalues of r lying between -1 and $+1$. This in turn is equivalent to $-c < r < c$. $\qquad\square$

A domain G in a complex vector space is called a *bounded symmetric domain* (in the sense of É. Cartan [Car]) if

(S.1) G is bounded.

(S.2) The group $H(G)$ of all biholomorphic automorphisms of G acts transitively on G.

(S.3) Given $a \in G$ there is an element $f \in H(G)$ such that $f^2 = \mathrm{Id}$ and a is an isolated fixed point of f.

We conclude from (S.2) that (S.3) is equivalent to the property that for at least one point in G there exists such a map f.

The domain $\tilde{Z}$ we are discussing is a bounded symmetric domain. Indeed, $\tilde{Z}$ is bounded, the group of biholomorphic automorphisms acts transitively, and $z \mapsto -z$ is an involution that has 0 as its only fixed point.

§7. Remarks on classification

Let us denote by **R** resp. **C** the field of real resp. complex numbers, by **H** the skew-field of (standard) Hamiltonian quaternions, and by **O** the alternative algebra of (standard) Cayley–Dickson numbers (octonions). These four algebras over **R** are of dimension 1, 2, 4, 8 respectively. Let **K** denote one of them. In **K** there is a standard involution $a \mapsto \bar{a}$, which fixes exactly the elements of **R** $\subset$ **K**. We consider the set $M^m(\mathbf{K})$ of all $m \times m$ matrices

$x = (\xi_{\nu\mu})$ with entries in $\mathbf{K}$. Hence $M^m(\mathbf{K})$ is a vector space over $\mathbf{R}$ of dimension $m^2, 2m^2, 4m^2, 8m^2$ respectively. On $M^m(\mathbf{K})$ we define an involution $x \mapsto \bar{x}^t$ by $\bar{x}^t = (\overline{\xi_{\mu\nu}})$ if $x = (\xi_{\nu\mu})$. Let

$$\mathcal{H}^m(\mathbf{K}) := \{x \in M^m(\mathbf{K}); \; \bar{x}^t = x\}.$$

Then this is a vector space over $\mathbf{R}$. We denote by xy the usual matrix product in $M^m(\mathbf{K})$ and put $x \circ y = \frac{1}{2}(xy + yx)$. Then $\mathcal{H}^m(\mathbf{K})$ is a formally real Jordan algebra over $\mathbf{R}$ if we assume $m \leq 3$ in the case $\mathbf{K} = \mathbf{O}$. For $\mathbf{K} \neq \mathbf{O}$ this is clear, only the case $\mathbf{K} = \mathbf{O}$ and $m = 3$ requires a longer calculation.

In Chapter III, §6, we saw that there is a Jordan algebra $\mathcal{A}$ associated with a given bilinear form μ. Let us denote the Jordan algebra in $\mathbf{R}^n$ given by the bilinear form

$$\mu(x,y) = x_1 y_1 - x_2 y_2 - \ldots - x_n y_n, \quad c = (1, 0, \ldots, 0),$$

by $\mathcal{A}_n$. Then $\mathcal{A}_n$ is formally real.

Now let us consider the following formally real Jordan algebras:

		dimension over $\mathbf{R}$ =	
$\mathbf{R}$			1
$\mathcal{A}_m$	, $m \geq 3$,		m
$\mathcal{H}^m(\mathbf{R})$	, $m \geq 3$,		$m(m+1)/2$
$\mathcal{H}^m(\mathbf{C})$	, $m \geq 3$,		m^2
$\mathcal{H}^m(\mathbf{H})$	, $m \geq 3$,		$2m^2 - m$
$\mathcal{H}^3(\mathbf{O})$	,		27

A classification theorem of P. Jordan, J. von Neumann and E. Wigner (cf. [15]) shows that these Jordan algebras are mutually non-isomorphic and that every simple formally real Jordan algebra is isomorphic to one of these. Hence the Decomposition Theorem III.11 shows that every semisimple formally real Jordan algebra is a direct sum of some copies of the algebras listed above. Each of these gives rise to a half-space as described in this chapter.

§8. One typical example

In this section we consider the set X of all $m \times m$ real symmetric matrices $x = (\xi_{kl})$. The dimension of the vector space X is $n = m(m+1)/2$. Let xy be the usual matrix product and $x \circ y = \frac{1}{2}(xy + yx)$. The algebra $\mathcal{A} = (X, \circ)$ is a Jordan algebra and $\mathcal{A}$ is formally real. The unit element of $\mathcal{A}$ is the unit matrix; we denote this matrix by e.

In this Jordan algebra two elements x, y commute if and only if $(xy - yx)z = z(xy - yz)$ holds for every $z \in X$. Since $xy - yx$ is a skew-symmetric matrix, it follows that x, y *commute in $\mathcal{A}$ if and only if $xy = yx$*. Hence the center of $\mathcal{A}$ consists of the elements λe, $\lambda \in \mathbf{R}$, and therefore $\mathcal{A}$ is a simple Jordan algebra. We also know that the inverse in $\mathcal{A}$ coincides with the inverse matrix.

We now prove that the bilinear form $\tau(x,y) = \mathrm{tr}\, L(x \circ y)$ can be described as

$$(1) \qquad\qquad \tau(x,y) = \frac{m+1}{2}\, \mathrm{tr}\,(xy).$$

Indeed, since $\sigma(x,y) = \mathrm{tr}\,(xy)$ is a symmetric bilinear form on $\mathcal{A}$ satisfying $\sigma(x \circ y, z) = \sigma(x, y \circ z)$, we conclude from Theorem III.10 that $\sigma(x,y) = \tau(d \circ x, y)$, where d belongs to the center of $\mathcal{A}$. Hence $\sigma(x,y) = \lambda \cdot \tau(x,y)$ for some λ, and $x = y = e$ yields (1).

For every real matrix r we define a linear transformation $W(r)$ on X by $W(r)x = rxr^t$. In Chapter I, §5 (4) we proved

$$(2) \qquad\qquad |\det W(r)| = |\det r|^{m+1}$$

as well as the identities $W(rs) = W(r)W(s)$ and $W^*(r) = W(r^t)$. We also know that the quadratic representation of $\mathcal{A}$ is given by

$$(3) \qquad\qquad P(a) = W(a) \quad \text{if} \quad a \in \mathcal{A}.$$

Obviously a linear transformation $W(r)$ is an automorphism of $\mathcal{A}$ if and only if r is an orthogonal matrix.

Let us now consider the mutation $\mathcal{A}_f$ for $f \in \mathcal{A}$. The definition of the product in $\mathcal{A}_f$ yields

$$x \perp y = \frac{1}{2}(xfy + yfx).$$

In view of $x \perp x = xfx$ the spectral theorem shows that $\mathcal{A}_f$ *is formally real if and only if f or $-f$ is positive definite.* We use this fact in order to prove that a linear transformation W of X satisfies

$$(4) \qquad\qquad W \in \Gamma(\mathcal{A}) \;\Leftrightarrow\; W = \pm V \cdot W(r),$$

where r is a non-singular real $m \times m$ matrix and V is an automorphism of the Jordan algebra $\mathcal{A}$.

Obviously, the linear transformations $\pm V \cdot W(r)$ belong to $\Gamma(\mathcal{A})$. Now let $W \in \Gamma(\mathcal{A})$ and consider $f = W^*e$. We know from Theorem IV.7 that the map $W : \mathcal{A}_f \to \mathcal{A}$ is an isomorphism. Since together with $\mathcal{A}$, its isomorphic images are formally real, we conclude that f or $-f$ is positive definite. Thus there exists a real matrix r, $\det r \neq 0$, such that $\pm e = W(r)f = W(r)W^*e$. Hence $V = \pm W(r)W^*$ belongs to $\Gamma(\mathcal{A})$ and satisfies $Ve = e$. Therefore V is an automorphism of $\mathcal{A}$.

The domain of positivity of $\mathcal{A}$ is given by

$$Y = \{y \in X;\ y\ \text{positive definite}\} \quad = \quad \{e(x);\ x \in X\},$$

thus the notion $y > 0$ in X means that y is positive definite. Let $\Sigma = \Sigma(Y)$ be the group of automorphisms of Y. Obviously, $W(r) \in \Sigma$ holds whenever $\det r \neq 0$.

Theorem 13. *(a) Every automorphism of the Jordan algebra $\mathcal{A}$ has the form $W(r)$, where r is some orthogonal matrix. The group $\mathrm{Aut}(\mathcal{A})$ is generated by the set*

$$\{P(r);\ r\ \text{symmetric and orthogonal}\}.$$

(b) $\Gamma(\mathcal{A})$ consists of all transformations $\pm W(r)$, where $\det r \neq 0$.

(c) Σ consists of all transformations $W(r)$, where $\det r \neq 0$.

Proof. (a) Let V be an automorphism of $\mathcal{A}$, hence $Ve = e$ and $V \in \Sigma$. Denote by e_k and $e_{kl}(k < l)$ the standard basis of the space of symmetric matrices. Since the e_k form an orthogonal system in $\mathcal{A}$, the Ve_k do so as well. The elements Ve_k of $\mathcal{A}$ commute pairwise and thus there is an orthogonal matrix r such that $W(r)Ve_k$ has diagonal form for all k. Without restriction we may assume that

$$Ue_k = e_k \quad \text{for every}\quad k, \quad \text{where}\quad U = W(r)V.$$

We need to show that the automorphism U is equal to $W(r')$ for some orthogonal r'.

Denote the vector space spanned by the e_k by D and denote by X' the set of all symmetric matrices in X for which the diagonal elements are zero. Then

$$X = D \oplus X'$$

and this sum is orthogonal with respect to $\tau(x, y)$. Moreover, we have for $x \in X$

(5) $$\tau(d, x) = 0 \quad \text{for every}\quad d \in D \quad \Rightarrow \quad x \in X'.$$

In view of $UD = D$ and $UU^* = \mathrm{Id}$ we obtain for $x \in X'$

$$\tau(d, Ux) = \tau(U^*d, x) = 0$$

for every $d \in D$. Hence we conclude from (5) that $UX' = X'$. For given $k < l$ we consider

$$y = \alpha e_k + \beta e_l + \gamma e_{kl} \in X.$$

Now we know

$$y \in \overline{Y} \quad \Leftrightarrow \quad \alpha\beta - \gamma^2 \geq 0, \; \alpha \geq 0, \; \beta \geq 0.$$

Moreover $y \in \overline{Y}$ implies

$$Uy = \alpha e_k + \beta e_l + \gamma U e_{kl} \in \overline{Y}.$$

In view of $U e_{kl} \in X'$ we conclude

$$U e_{kl} = \lambda_{kl} e_{kl} \quad \text{and} \quad |\lambda_{kl}| \leq 1.$$

The same argument for U^{-1} leads to $\lambda_{kl} = \pm 1$. If $k < l < m$ one has $e_{kl} \circ e_{lm} = \frac{1}{2} e_{km}$, hence $\lambda_{kl} \lambda_{lm} = \lambda_{km}$. Hence we obtain $U = W(r')$ for a diagonal orthogonal matrix r'.

Thus every automorphism of $\mathcal{A}$ can be written in the form $W(r)$, where r is an orthogonal matrix. Since every orthogonal matrix r has the form

$$r = r_1 \quad \text{or} \quad r = r_1 \cdot s$$

where r_1 belongs to the connected component of the identity, and where $s = -e_1 + e_2 + \ldots + e_m$ we conclude that every automorphism V of $\mathcal{A}$ has the form

$$V = V_1 \quad \text{or} \quad V = V_1 \cdot W(s),$$

where V_1 belongs to the connected component of the identity in the group of automorphisms. Hence Theorem IV.9 shows that V_1 is a product of transformations $P(r)$ for symmetric orthogonal matrices r. Since s is also symmetric we get $W(s) = P(s)$ and this proves part (a).

(b) The claim follows immediately from (4) and (a).

(c) Note that $\Sigma \subset \Gamma(\mathcal{A})$. $\qquad\qquad\qquad\qquad\qquad\qquad\qquad\qquad\square$

Now we consider the half-space associated with $\mathcal{A}$:

$$Z = \{z = x + iy; \; x, y \text{ real symmetric } m \times m \text{ matrices}, \; y > 0\}.$$

This domain is called the *Siegel half-space*. Theorems 11 and 13 lead to

Theorem 14. *The group of biholomorphic automorphisms of the Siegel half-space is generated by the mappings*

$$z \mapsto -z^{-1} \quad \text{and} \quad z \mapsto z + a,$$

with a real symmetric.

Let us now consider real matrices

$$M = \begin{pmatrix} a & b \\ c & d \end{pmatrix}$$

of type $2m \times 2m$, with a, b, c, d of type $m \times m$. Let

$$J = \begin{pmatrix} 0 & e \\ -e & 0 \end{pmatrix}$$

and consider the group S of *symplectic matrices*, i.e. those matrices M satisfying $M^t J M = J$. On the other hand we get for a complex symmetric matrix $z = x + iy$

(6)
$$-\frac{1}{2i} \begin{pmatrix} \bar{z} \\ e \end{pmatrix}^t J \begin{pmatrix} z \\ e \end{pmatrix} = y, \quad M \begin{pmatrix} z \\ e \end{pmatrix} = \begin{pmatrix} az + b \\ cz + d \end{pmatrix}.$$

Thus Z is given as the set of matrices z for which the left hand side of (6) is positive definite. We prove for $z \in Z$ and $M \in S$ that the matrix $cz + d$ has an inverse. Let v be a complex vector such that $(cz + d)v = 0$. Then we obtain $\bar{v}^t y v = 0$, using (6). In view of $y > 0$ it follows that $v = 0$.

Thus, for $z \in Z$ and $M \in S$ the mapping

$$z \mapsto Mz := (az + b)(cz + d)^{-1}$$

is defined and holomorphic in $z \in Z$. It is called a *symplectic transformation* The equation

$$\begin{pmatrix} Mz \\ e \end{pmatrix} = M \begin{pmatrix} z \\ e \end{pmatrix} (cz + d)^{-1}$$

yields

$$\mathrm{Im}\,(Mz) = \frac{-1}{2i} \left(\overline{\frac{Mz}{e}}\right)^t J \begin{pmatrix} Mz \\ e \end{pmatrix} = \overline{(cz + d)^{-1}}^t \cdot (\mathrm{Im}\,(z)) \cdot (cz + d)^{-1}$$

and consequently $Mz \in Z$ for $z \in Z$.

An elementary calculation leads to

$$M_1(M_2 z) = (M_1 M_2)z, \quad M_k \in S,$$

and hence $z \mapsto Mz$, $M \in S$, is a biholomorphic automorphism of Z.

Considering $M = \begin{pmatrix} e & b \\ 0 & e \end{pmatrix}$, b real symmetric, and $M = \begin{pmatrix} 0 & e \\ -e & 0 \end{pmatrix}$ we see that the maps $z \mapsto z + b$ and $z \mapsto -z^{-1}$ are symplectic.

Thus Theorem 14 leads to

Theorem 15. *The group of biholomorphic automorphisms of the Siegel half-space is equal to the group of symplectic transformations.*

Since $M_1 z = M_2 z$ for every $z \in Z$ implies $M_1 = \pm M_2$, we obtain

Corollary 3. *The group S of sympletic matrices is generated by the matrices*

$$\begin{pmatrix} 0 & e \\ -e & 0 \end{pmatrix} \quad and \quad \begin{pmatrix} e & b \\ 0 & e \end{pmatrix},$$

with b real symmetric.

Notes

For the half-spaces associated with formally real Jordan algebras U. Hirze-bruch [5] investigated the group of biholomorphic automorphisms and obtained the result that this group is generated as described in Theorem 11(b). Some results in this case are also due to O.S. Rothaus [21] and A. Korányi [19].

The half-space associated with the formally real Jordan algebra of real symmetric matrices was investigated by C.L. Siegel [25].

Editors' Notes

1. (i) The results on holomorphic functions used in §1 are fairly standard and can be found for instance in Range [Ra].

(ii) The set H introduced in §1 should actually be viewed as a presheaf. As the results of sections 1, 2 and 3 show, H may be identified with a group of birational transformations of $X(i)$. It coincides with $H(Z)$ if $\mathcal{A}$ is formally real.

(iii) The mapping p in §2(1) is usually called the *generalized Cayley transformation*. It was investigated by U. Hirzebruch [5] as well as by Korányi and Wolf [KW].

(iv) The identity $t_a \cdot j \cdot t_{a^{-1}} \cdot j \cdot t_a \cdot j = P(a)$ in Lemma 7 is usually called the *Hua identity*. It is known that this identity, in turn, characterizes inverses in Jordan algebras; see Springer [Spr], McCrimmon [McC1], and [GrW]. The identity is also important for computational purposes; for instance it has been used in [KrW] to determine factors of automorphy on half-spaces.

(v) If one considers on $H(Z)$ the topology arising from uniform convergence on compact subsets of Z then $H(Z)$ becomes a topological group, even a Lie group. The associated Lie algebra of $\mathcal{A}$ is canonically isomorphic to the Kantor–Koecher–Tits algebra; cf. [Ka], [Ko1], [Ti], [KrP].

(vi) Just as in the case of formally real Jordan algebras by U. Hirzebruch [5], fundamental results of the Bergman theory (cf. [B]) led to the proof of Theorems 5 and 7. In the case of the Siegel half-space as considered in §8 however, Siegel [25] used a generalized version of Schwarz's Lemma for his proof of Theorem 15.

(vii) The group of automorphisms of the half-space associated with a circular cone (cf. Chapter I, §5 (C)), which corresponds to a simple formally real Jordan algebra of type $\mathcal{A}_m$, was described in detail by Helwig [H1].

2. In §6, Koecher constructed a bounded symmetric domain in $\mathbf{C}^n$ in the sense of Cartan [Car] for each semisimple formally real Jordan algebra via the generalized Cayley transformation. The Jordan theoretic approach to bounded symmetric domains is studied in detail in Koecher's Rice Notes [Ko2]. For the general setting one needs the broader concept of Jordan triple systems. Loos [Lo3] used this general method in order to describe a correspondence between positive hermitian Jordan triple systems and all the bounded symmetric domains appearing in the Cartan classification (cf. [He], VII §7). The basic advantage is that one can replace Lie theory by the more elementary techniques of Jordan theory in this construction. Moreover Kaup ([Ka1]–[Ka3], [BKU], [KU]) was able to classify the infinite dimensional bounded symmetric domains by using a suitable extension of the Jordan approach. It is worth mentioning that in the infinite dimensional case there is no counterpart to Cartan's approach via Lie groups.

The half-spaces considered in this chapter give an essential part of the bounded symmetric domains or Hermitian symmetric spaces. They are also called *Siegel domains* of the first kind. Pyatetskii–Shapiro (cf. [P-S]) found a generalization of half-spaces called Siegel domains of the second kind which allow to realize all symmetric domains in such a form. On the algebraic level this corresponds to the transition from Jordan algebras to Jordan triple systems. Furthermore Dorfmeister [Do5], [Do6] and Satake [Sa1], [Sa2] investigated the special case of quasisymmetric domains. For the automorphisms of Siegel domains, see also Kaup, Matsushuina and Ochiai [KMO], Murakami [Mu], and Satake [Sa3].

Moreover the construction of Siegel domains of the second kind enabled Pyatetskii-Shapiro to give an example of a bounded homogeneous non-symmetric domain in $\mathbf{C}^n$. Later Vinberg, Gindikin and Pyatetskii-Shapiro

[VGP-S] proved that such domains possess realizations as Siegel domains of the second kind.

In recent years also half-spaces have been considered which correspond to non-convex cones. They play an essential role in the realization of pseudo-Hermitian symmetric spaces, confer D'Atri and Gindikin [AGi] as well as Faraut and Gindikin [FGi].

In [17] Koecher had initiated a theory of automorphic forms on half-spaces. There has been a lot of progress since the completion of the original "Minnesota Notes" (cf. [Kr1]). In particular this concerns the work of Resnikoff [Re] and Dorfmeister [Do2] on theta series and the papers on Eisenstein and theta series on the exceptional 27–dimensional half-space associated with $\mathcal{H}^3(\mathbf{O})$ (cf. [Ba], [Ki], [Kr2]). Here Koecher's approach using Jordan theory has proved to be extremely efficient.

Finally, recent work using Jordan algebras in the investigation of group representations and associated analytic objects should be mentioned. See, for instance, Kostant and Sahi [KsSi], or Bertram and Hilgert [BeHi].

Appendix: The Bergman kernel function

§1. Reproducing kernels

Let Z be a non-empty set and let $\mathcal{H}$ be a vector space over $\mathbf{C}$ of complex-valued functions on Z. A map $\mathcal{H} \times \mathcal{H} \to \mathbf{C}$, denoted by (f,g), is called a *positive definite hermitian form* on $\mathcal{H}$ if

(H.1) $\overline{(f,g)} = (g,f)$.

(H.2) (f,g) is linear in the first argument.

(H.3) $(f,f) \geq 0$ and $(f,f) = 0$ only if $f = 0$.

As usual we get a *norm* $|f|$ on $\mathcal{H}$ by $|f| := \sqrt{(f,f)}$, satisfying the properties

$$|\alpha f| = |\alpha| \cdot |f|, \quad |f + g| \leq |f| + |g|, \quad |f| > 0 \quad \text{if} \quad f \neq 0, \quad |0| = 0$$

and $|(f,g)| \leq |f| \cdot |g|$. Since $|f - g|$ is a metric on $\mathcal{H}$, the vector space is a topological vector space.

Let $k_y : Z \to \mathbf{C}$ be a function for every $y \in Z$. Then k_y is called a *reproducing kernel of* $\mathcal{H}$ if

(R.1) For every $y \in Z$ the function k_y belongs to $\mathcal{H}$.

(R.2) For every $y \in Z$ and $f \in \mathcal{H}$ one has $f(y) = (f, k_y)$.

If such a reproducing kernel k_y exists, then it is uniquely determined. Indeed, let k_y' be another reproducing kernel of $\mathcal{H}$, then

$$|k_y - k_y'|^2 = (k_y - k_y', \ k_y) - (k_y - k_y', \ k_y') = 0$$

and hence $k_y' = k_y$.

Let k_y be the reproducing kernel of $\mathcal{H}$, then set

(1) $$k(y) := \sqrt{(k_y, k_y)} = |k_y| = k_y(y),$$

and we immediately obtain

(2) $$f(y) \leq k(y) \cdot |f| \quad \text{for} \quad f \in \mathcal{H}.$$

Now let us assume that $\mathcal{H}$ is a *Hilbert space*, i.e. that there is a positive definite hermitian form on $\mathcal{H}$ and that $\mathcal{H}$ is complete with respect to the corresponding norm topology.

Lemma 1. *Let Φ be a continuous linear form on a complex Hilbert space $\mathcal{H}$. Then there exists a uniquely determined element $g \in \mathcal{H}$ such that*

$$\Phi(f) = (f, g) \quad \text{for every} \quad f \in \mathcal{H}.$$

Using Lemma 1 it is easy to prove

Theorem 1. *Let $\mathcal{H}$ be a Hilbert space of complex valued functions on a non-empty set Z. There exists a reproducing kernel of $\mathcal{H}$ if and only if there is a non-negative function ϱ on Z such that*

$$|f(x)| \leq \varrho(x) \cdot |f| \quad \text{for every} \quad f \in \mathcal{H} \quad \text{and} \quad x \in Z.$$

Proof. If there is a reproducing kernel of $\mathcal{H}$, then formula (2) shows the existence of ϱ. Conversely, for given $y \in Z$ let $\Phi(f) = f(y)$. Then Φ is a linear form on $\mathcal{H}$ and the hypothesis states that Φ is continuous. Lemma 1 yields $k_y \in \mathcal{H}$ such that $f(y) = (f, k_y)$. Hence k_y is a reproducing kernel of $\mathcal{H}$.   □

Without proof we mention

Theorem 2. *Let $\mathcal{H}$ be a Hilbert space of complex-valued functions on a non-empty set Z. Then the existence of a reproducing kernel for $\mathcal{H}$ is equivalent to the following: Every orthogonal system $\varphi_1, \varphi_2, \ldots$ in $\mathcal{H}$ satisfies*

(a) $f(x) = \Sigma_k \alpha_k \varphi_k(x)$ for every $x \in Z$ in the sense of absolute convergence if $f = \Sigma_k \alpha_k \varphi_k$,

(b) $\Sigma_k |\varphi_k(x)|^2$ is convergent.

In order to apply our results we consider a group Ω of bijective mappings of Z onto itself. A map $\Omega \times Z \to \mathbf{C}$, $(\alpha, x) \mapsto j_\alpha(x)$, is called a *factor of automorphy* (of Ω) if

$$j_{\alpha\beta}(x) = j_\alpha(\beta x) j_\beta(x), \quad \alpha, \beta \in \Omega,$$

holds identically in x. For every function $f : Z \to \mathbf{C}$ and for every factor of automorphy j we set

$$f^\alpha(x) = j_\alpha(x) \cdot f(\alpha x).$$

Obviously we have $(f^\alpha)^\beta = f^{\alpha\beta}$ for $\alpha, \beta \in \Omega$. Now let $\Gamma = \Gamma(\mathcal{H}, \Omega, j)$ be the group of $\alpha \in \Omega$ such that

(i) the mapping $f \mapsto f^\alpha$ maps $\mathcal{H}$ onto $\mathcal{H}$,

(ii) $(f^\alpha, g^\alpha) = (f, g)$ for every $f, g \in \mathcal{H}$.

Theorem 3. *Let* $\mathcal{H} \neq \{0\}$ *be a Hilbert space of complex-valued functions on a non-empty set* Z. *If there is a non-negative function* $\varrho(x)$ *such that* $|f(x)| \leq \varrho(x) \cdot |f|$ *for every* $f \in \mathcal{H}$ *and* $x \in Z$, *then there exists a function* $k : Z \to \mathbf{C}$ *such that* k *does not vanish identically and*

$$|j_\alpha(x)|^2 \cdot k(\alpha x) = k(x) \quad for \quad x \in Z \quad and \quad \alpha \in \Gamma.$$

In this context $k(x)$ is called the *Bergman kernel*.

Proof. Theorem 1 shows that there is a reproducing kernel k_y of $\mathcal{H}$ which is not zero. Given $\alpha \in \Gamma$ it follows that

$$f^\alpha(y) = j_\alpha(y)f(\alpha y) = j_\alpha(y)(f, k_{\alpha y}) = j_\alpha(y)(f^\alpha, k_{\alpha y}^\alpha) = (f^\alpha, k_y^*),$$

where

$$k_y^*(x) = j_\alpha(x)\overline{j_\alpha(y)}k_{\alpha y}(\alpha x).$$

Hence k_y^* is a reproducing kernel and therefore $k_y^* = k_y$ due to uniqueness. Putting $k(x) = k_x(x)$ we get the claim. $\qquad\square$

§2. Domains in complex number space

Let $\mathbf{C}^n$ be the vector space of the n–tuples $z = (z_1, \ldots, z_n)$ of complex numbers z_k and let Z be a non-empty open subset of $\mathbf{C}^n$. We denote by $\mathrm{Hol}(Z)$ the vector space of all holomorphic functions $f : Z \to \mathbf{C}$ such that the integral

$$|f|^2 = |f|_Z^2 = \int_Z |f(z)|^2 dx\, dy, \quad z = x + iy,$$

exists. Putting

$$(f, g) = \int_Z f(z)\overline{g(z)}dx\, dy$$

for $f, g \in \mathrm{Hol}(Z)$ we see that (f, g) is a positive definite hermitian form on $\mathrm{Hol}(Z)$ and that $\mathrm{Hol}(Z)$ is a metric space with the norm topology induced by (f, g).

Given $a \in Z$ we denote by

$$\mu(a) = \inf\{|z - a|; \quad z \in \partial Z\}, \quad |z|^2 = \bar{z}^t z,$$

the distance of a to the boundary of Z, where $\mu(a)$ can be chosen as an arbitrary positive number in case $Z = \mathbf{C}^n$. Moreover, put

$$\varrho(a) = \left(\frac{n}{\pi}\right)^{n/2} \cdot \mu^{-n}(a).$$

Lemma 2. *Given $f \in \mathrm{Hol}(Z)$ and $a \in Z$ one has*

$$|f(a)| \leq \varrho(a) \cdot |f|.$$

Proof. For positive μ let P_μ be the polycylinder with center a and radius μ, i.e.

$$P_\mu := \{z \in \mathbf{C}^n;\ |z_k - a_k| < \mu \quad \text{for} \quad k = 1, \ldots, n\}.$$

Hence $P_\mu \subset Z$ holds provided that $\mu < \frac{1}{\sqrt{n}}\mu(a)$. Since $f(z)$ is holomorphic on Z, there exists a power series expansion

$$f(z) = \sum_g \alpha_g (z - a)^g, \quad z \in P_\mu,$$

where g runs through all the n–tuples whose components are non-negative integers and

$$(z - a)^g := \prod_k (z_k - a_k)^{g_k},$$

For $\mu < \frac{1}{\sqrt{n}}\mu(a)$ we get

$$|f|^2 \geq \int_{P_\mu} |f(z)|^2 dx\, dy = \sum_{g,h} \alpha_g \bar{\alpha}_h \int_{P_\mu} (z - a)^g \overline{(z - a)^h} dx\, dy.$$

Using polar coordinates in the components of $z - a$ it is easily seen that the integrals on the right hand side vanish for $g \neq h$. Hence

$$|f|^2 \geq |\alpha_0|^2 \int_{P_\mu} dx\, dy = (\pi \cdot \mu^2)^n \cdot |f(a)|^2.$$

For $\mu \to \frac{1}{\sqrt{n}}\mu(a)$ the claim of the lemma follows. $\square$

Corollary 1. *Every $f \in \mathrm{Hol}(\mathbf{C}^n)$ vanishes identically.*

In order to prove that $\mathrm{Hol}(Z)$ is a Hilbert space we only need to show that $\mathrm{Hol}(Z)$ is complete in the norm topology. Let $(f_p)_p$ be a Cauchy sequence. Then Lemma 2 shows that for every compact set $K \subset Z$ there is a positive number $\varrho(K)$ such that

$$|f_p(z) - f_q(z)| \leq \varrho(K) \cdot |f_p - f_q| \quad \text{for} \quad z \in K.$$

Hence $(f_q(z))_q$ converges uniformly for $z \in K$ to a holomorphic function $f(z)$ on Z. Moreover, we get

$$|f|_K^2 = \lim_{q \to \infty} |f_q|_K^2 \leq \sup\{|f_q|^2;\ q \geq 1\}.$$

Hence $f \in \mathrm{Hol}(Z)$ follows. One can easily prove that $(f_q)_q$ converges to f in $\mathrm{Hol}(Z)$. Therefore $\mathrm{Hol}(Z)$ is a Hilbert space.

Theorem 1 states that $\mathrm{Hol}(Z)$ possesses a reproducing kernel.

Denote by Ω the group of all biholomorphic automorphisms of Z and by

$$j_\alpha(z) = \det \frac{\partial \alpha(z)}{\partial z} \quad \text{for} \quad \alpha \in \Omega$$

the Jacobian of the map $z \mapsto \alpha(z)$. Obviously, $j_\alpha(z)$ is a factor of automorphy, and the definition of (f, g) shows $(f^\alpha, g^\alpha) = (f, g)$. Hence $f \mapsto f^\alpha$ maps $\mathrm{Hol}(Z)$ onto itself.

Theorem 3 now yields the existence of the Bergman kernel:

Theorem 4. *Let $Z \neq \mathbf{C}^n$ be an open subset of $\mathbf{C}^n$ such that there is at least one not-identically vanishing holomorphic function on Z with $|f| < \infty$. Then there exists a function $k(z)$ on Z such that k does not vanish identically and*

$$\left| \det \frac{\partial \alpha(z)}{\partial z} \right|^2 k(\alpha(z)) = k(z), \quad z \in Z,$$

for every biholomorphic automorphism α of Z.

Notes

For the theory of the Bergman kernel function see N. Aronszajn [2] and the literature cited there.

Editors' Notes

1.(i) Lemma 1 is the well-known Riesz representation theorem.

(ii) Theorem 4 in particular guarantees the existence of a Bergman kernel for bounded domains in $\mathbf{C}^n$. Using the map $\tilde{p}$ in Chapter VII, §2 (2) we also obtain a Bergman kernel on the half-space Z associated with a semisimple formally real Jordan algebra.

2. The results of this appendix are also contained e.g. in the books by Baily [Ba], Faraut and Korányi [FKo], Range [Ra], and Satake [Sa2], Chap.II.

References

[0] E. Artin, H. Braun and M. Koecher: Unpublished manuscript.

[1] A.A. Albert: A structure theory for Jordan algebras. Ann. Math. **48** (1947), 546–567.

[2] N. Aronszajn: Theory of reproducing kernels. Trans. Amer. Math. Soc. **68** (1950), 337–404.

[3] B. Harris: Centralizers in Jordan algebras. Pacific J. Math. **8** (1958), 757–790.

[4] Ch. Hertneck: Positivitätsbereiche und Jordan–Strukturen. Math. Ann. **146** (1962), 433–455.

[5] U. Hirzebruch: Halbräume und ihre holomorphen Automorphismen. Math. Ann. **153** (1964), 395–417.

[6] N. Jacobson: The center of a Jordan ring. Bull. Amer. Math. Soc. **54** (1948), 316–322.

[7] N. Jacobson: Operator commutativity in Jordan algebras. Proc. Amer. Math. Soc. **3** (1952), 973–976.

[8] N. Jacobson: A theorem on the structure of Jordan algebras. Proc. Natl. Acad. Sci. USA **42** (1956), 140–147.

[9] N. Jacobson: Jordan algebras. In: Report of a conference on linear algebras. Natl. Research Council, Publ. **502** (1957), 12–19.

[10] N. Jacobson: A coordinatization theorem for Jordan algebras. Proc. Natl. Acad. Sci. USA **48** (1962), 1154–1160.

[11] N. Jacobson: Some groups of transformations defined by Jordan algebras I, II, III. J. Reine Angew. Math. **201** (1959), 178–195; **204** (1960), 74–98; **207** (1961), 61–85.

[12] N. Jacobson: Macdonald's theorem on Jordan algebras. Archiv Math. **13** (1962), 241–250.

[13] P. Jordan: Über eine Klasse nicht assoziativer hyperkomplexer Algebren. Nachr. Ges. Wiss. Göttingen (1932), 569–575.

[14] P. Jordan: Über Verallgemeinerungsmöglichkeiten des Formalismus der Quantenmechanik. Nachr. Ges. Wiss. Göttingen (1933), 209–217.

[15] P. Jordan, J. von Neumann, E. Wigner: On an algebraic generalization of the quantum mechanical formalism. Ann. Math. **35** (1934), 29–64.

[16] M. Koecher: Positivitätsbereiche im $\mathbf{R}^n$. Amer. J. Math. **79** (1957), 575–596.

[17] M. Koecher: Automorphic forms in half-spaces. Seminar on analytic functions, Institute for Advanced Study, Princeton, Vol. **2** (1958), 105–119.

[18] M. Koecher: Die Geodätischen von Positivitätsbereichen. Math. Ann. **135** (1958), 192–202.

[19] M. Koecher: Eine Charakterisierung der Jordan–Algebren. Math. Ann. **148** (1962), 244–256.

[20] A. Korányi: Lecture notes. University of California, Berkeley (1961).

[21] I.G. Macdonald: Jordan algebras with three generators. Proc. London Math. Soc. **10** (1960), 395–408.

[22] O.S. Rothaus: Domains of positivity. Abh. Math. Sem. Univ. Hamburg **24** (1960), 189–235.

[23] R.D. Schafer: An introduction to nonassociative algebras. Academic Press, New York (1966).

[24] R.D. Schafer: Structure and representation of nonassociative algebras. Bull. Amer. Math. Soc. **61** (1955), 469–484.

[25] C.L. Siegel: Symplectic geometry. Amer. J. Math. **65** (1943), 1–86; Ges. Abhandlungen II, 274–359.

[26] E.B. Vinberg: Homogeneous cones. Sov. Math. Dokl. **1** (1960), 787–790.

Editors' References

[AGi] J.E. D'Atri, S. Gindikin: Siegel domain realization of pseudo-Hermitian symmetric manifolds. Geom. Dedicata **46** (1993), 91–125.

[Ba] W.L. Baily: Introductory lectures on automorphic forms. Iwanami Shoten and Princeton University Press, Tokyo–Princeton (1973).

[B] S. Bergman: The kernel function and conformal mapping. Math. Surveys V, Amer. Math. Soc., New York (1950).

[Be] E. Bernadac: Random continued fractions and inverse Gaussian distribution on a symmetric cone. J. Theoret. Prob. **8** (1995), 221–259.

[BeHi] W. Bertram, J. Hilgert: Hardy spaces and analytic continuation of Bergman spaces. Bull. Soc. Math. France **126** (1998), 435–482.

[BK] H. Braun, M. Koecher: Jordan–Algebren. Springer, Berlin-Heidelberg-New York (1966).

[BKU] R. Braun, W. Kaup, H. Upmeier: A holomorphic characterization of Jordan C^*-algebras. Math. Z. **161** (1978), 277–290.

[Car] É. Cartan: Sur les domaines bornés homogènes de l'espace de n variables. Abh. Math. Semin. Univ. Hamburg **11** (1935), 116–162.

[Cas] M. Casalis: Les familles exponentielles à variance quadratique homogène sont les lois de Wishart sur une cône symétrique. C. R. Acad. Sci. Paris **312** (1991), 537–540.

[CL] M. Casalis, G. Letac: The Lukacs-Olkin-Rubin characterization of Wishart distributions on symmetric cones. Ann. Statist. **24** (1996), 763–786.

[Do1] J. Dorfmeister: Zur Konstruktion homogener Kegel. Math. Ann. **216** (1975), 79–96.

[Do2] J. Dorfmeister: Theta functions for special, formally real Jordan algebras. (A remark on a paper of H.L. Resnikoff.) Invent. Math. **44** (1978), 103–108.

[Do3] J. Dorfmeister: Inductive construction of homogeneous cones. Trans. Amer. Math. Soc. **252** (1979), 321–349.

[Do4] J. Dorfmeister: Algebraic description of homogeneous cones. Trans. Amer. Math. Soc. **255** (1979), 61–89.

[Do5] J. Dorfmeister: Quasisymmetric Siegel domains and the automorphisms of homogeneous Siegel domains. Amer. J. Math. **102** (1980), 537–563.

[Do6] J. Dorfmeister: Homogeneous Siegel domains. Nagoya Math. J. **86** (1982), 39–83.

[DK1] J. Dorfmeister, M. Koecher: Relative Invarianten und nichtassoziative Algebren. Math. Ann. **228** (1977), 147–186.

[DK2] J. Dorfmeister, M. Koecher: Reguläre Kegel. Jahresber. Deutsche Math.-Verein. **81** (1979), 109–151.

[FGi] J. Faraut, S. Gindikin: Pseudo-hermitian symmetric spaces of tube type. In: Topics in Geometry (S. Gindikin, ed.), Birkhäuser, Boston (1996), 123–154.

[FKo] J. Faraut, A. Korányi: Analysis on symmetric cones. Clarendon Press, Oxford (1994).

[F-L] A. Fernandez-Lopez (ed.): Actas de la conferencia internacional sobre estructuras de Jordan. Servicio de Publicaciones de Universidad de Malaga, Malaga (to appear).

[Gi1] S. Gindikin: Analysis on homogeneous domains. Russian Math. Surveys **19** (4) (1964), 1-89.

[Gi2] S. Gindikin: Fourier transform and Hardy spaces of $\bar{\partial}$-cohomology in tube domains. C. R. Acad. Sci. Paris **315** (1992), 1139–1143.

[Gi3] S. Gindikin: Holomorphic language for $\bar{\partial}$-cohomology and representations of real semisimple Lie groups. Contemp. Math. **154** (1993), 103–115.

[GrW] H. Gradl, S. Walcher: On a class of inversions. Comm. Algebra **20** (1992), 2371–2392.

[HU] U. Hagenbach, H. Upmeier: Toeplitz C^*-algebras over non-convex cones and pseudo-symmetric spaces. Contemp. Math. **212** (1998), 109–131.

[H-OS] H. Hanche-Olsen, E. Størmer: Jordan operator algebras. Pitman, London (1984).

[He] S. Helgason: Differential geometry, Lie groups and symmetric spaces. Academic Press, New York (1978).

[H1] K.-H. Helwig: Automorphismengruppen des allgemeinen Kreiskegels und des zugehörigen Halbraumes. Math. Ann. **157** (1964), 1–33.

[H2] K.-H. Helwig: Jordan–Algebren und symmetrische Räume. I. Math. Z. **115** (1970), 315–349.

[Hir] U. Hirzebruch: Über Jordan–Algebren und beschränkte symmetrische Gebiete. Math. Z. **94** (1966), 387–390.

[Jac1] N. Jacobson: Derivation algebras and multiplication algebras of semisimple Jordan algebras. Ann. Math. **50** (1949), 866–874.

[Jac2] N. Jacobson: Structure and representations of Jordan algebras. Amer. Math. Soc. Colloquium Publ., Providence, RI (1968).

[Jos] J. Jost: Riemannian geometry and geometric analysis. Springer, Berlin–Heidelberg–New York (1995).

[Ka] I.L. Kantor: Classification of irreducible transitively differential groups. Soviet Math. Dokl. **5** (1965), 1404–1407.

[Ka1] W. Kaup: Algebraic characterization of symmetric complex Banach manifolds. Math. Ann. **228** (1977), 39–64.

[Ka2] W. Kaup: Über die Klassifikation der symmetrischen hermiteschen Mannigfaltigkeiten unendlicher Dimension I, II. Math. Ann. **257** (1981), 463–483; **262** (1983), 57–75.

[Ka3] W. Kaup: A Riemann mapping theorem for bounded symmetric domains in complex Banach spaces. Math. Z. **183** (1983), 503–529.

[KMO] W. Kaup, Y. Matsushima, T. Ochiai: On the automorphisms and equivalences of generalized Siegel domains. Amer. J. Math. **92** (1970), 475–498.

[KMP] W. Kaup, K. McCrimmon, H. Petersson (eds.): Jordan algebras. Proceedings of a conference held at Oberwolfach 1992. Walter de Gruyter, Berlin (1994).

[KU] W. Kaup, H. Upmeier: Jordan algebras and symmetric Siegel domains in Banach spaces. Math. Z. **157** (1977), 179–200.

[Ki] H. Kim: Exceptional modular form of weight 4 on an exceptional domain contained in $\mathbf{C}^{27}$. Rev. Mat. Iberoamericana **9** (1993), 139–200.

[Ko1] M. Koecher: Imbedding of Jordan algebras into Lie algebras I, II. Amer. J. Math. **89** (1967), 787–816; **90** (1968), 476–510.

[Ko2] M. Koecher: An elementary approach to bounded symmetric domains. Rice University, Houston (1969), 1-143.

[Ko3] M. Koecher: Lineare Algebra und analytische Geometrie. 4th ed., Springer, Berlin–Heidelberg–New York (1997).

[KW] A. Korányi, J.A. Wolf: Realization of Hermitian symmetric spaces as generalized half–planes. Ann. Math. **81** (1965), 265–288.

[KsSi] B. Kostant, S. Sahi: Jordan algebras and Capelli identities. Invent. Math. **112** (1993), 657–664.

[Kr1] A. Krieg: Modular forms on half–spaces of quaternions. Lect. Notes Math. **1143**, Springer, Berlin–Heidelberg–New York (1985).

[Kr2] A. Krieg: The singular modular forms on the 27–dimensional exceptional domain. Manuscripta Math. **92** (1997), 361–367.

[KrP] A. Krieg, H. Petersson: Max Koecher zum Gedächtnis. Jahresber. Deutsche Math.-Verein. **95** (1993), 1–27.

[KrW] A. Krieg, S. Walcher: Multiplier systems for the modular group on the 27–dimensional exceptional domain. Comm. Algebra **26** (1998), 1409–1417.

[LM] G. Letac, H. Massam: Craig–Sakamoto's theorem for the Wishart distribution on symmetric cones. Ann. Inst. Stat. Math. **47** (1995), 785–799.

[Lo1] O. Loos: Symmetric spaces I, II. Benjamin, Reading, MA (1969).

[Lo2] O. Loos: Jordan pairs. Lect. Notes Math. **460**, Springer, Berlin-Heidelberg-New York (1975).

[Lo3] O. Loos: Bounded symmetric domains. University of California Lecture Notes, Irvine (1977).

[MNe1] H. Massam, E. Neher: On transformations and determinants of Wishart variables on symmetric cones. J. Theoret. Prob. **10** (1997), 867–902.

[MNe2] H. Massam, E. Neher: Estimation and testing for lattice conditional independence models on Euclidean Jordan algebras. Ann. Statist. **26** (1998), 1051–1082.

[McC1] K. McCrimmon: Axioms for inversion in Jordan algebras. J. Algebra **47** (1977), 201–222.

[McC2] K. McCrimmon: The Russian revolution in Jordan algebras. Algebras Groups Geom. **1** (1984), 1–61.

[Mey] K. Meyberg: Lectures on algebras and triple systems. University of Virginia Lecture Notes, Charlottesville (1972).

[MRS] I. Müller, H. Rubenthaler, G. Schiffmann: Structure des espaces préhomogènes associés à certaines algèbres de Lie graduées. Math. Ann. **274** (1986), 95–123.

[Mu] S. Murakami: On automorphisms of Siegel domains. Lect. Notes Math. **286**, Springer, Berlin-Heidelberg-New York (1972).

[Ne1] E. Neher: Transformation groups of the Andersson–Perlman cone. J. Lie Theory **9** (1999), 203–213.

[Ne2] E. Neher: A statistical model for Euclidean Jordan algebras. To appear in [F-L].

[P-S] I.-I. Pyatetskii-Shapiro: Automorphic functions and the geometry of classical domains. Gordon and Breach, New York–London–Paris (1969).

[Ra] R.M. Range: Holomorphic functions and integral representations in several complex variables. Springer, New York–Berlin–Heidelberg (1986).

[Re] H.L. Resnikoff: Theta functions for Jordan algebras. Invent. Math. **31** (1975), 87–104.

[Sa1] I. Satake: On classification of quasisymmetric domains. Nagoya Math. J. **62** (1976), 1–12.

[Sa2] I. Satake: Algebraic structures of symmetric domains. Iwanami Shoten and Princeton University Press, Tokyo–Princeton (1980).

[Sa3] I. Satake: On **Q**-structures of quasisymmetric domains. Tohoku Math. J. **47** (1995), 357–390; Addendum. 613–614.

[SKi] M. Sato, T. Kimura: A classification of irreducible prehomogeneous vector spaces and their relative invariants. Nagoya Math. J. **65** (1977), 1–155.

[Sh] H. Shima: On locally symmetric homogeneous domains of completely reducible linear Lie groups. Math. Ann. **217** (1975), 93–95.

[Spr] T.A. Springer: Jordan algebras and algebraic groups. Springer, Berlin–Heidelberg–New York (1973); Reprint (1998).

[Ti] J. Tits: Une classe d' algèbres de Lie en relation avec des algèbres de Jordan. Indag. Math. **24** (1962), 530–535.

[Vi] E.B. Vinberg: The theory of convex homogeneous cones. Trans. Moscow Math. Soc. **12** (1963), 340–403.

[VGP-S] E.B. Vinberg, S. Gindikin, I.-I. Pyatetskii-Shapiro: Classification and canonical realization of complex bounded homogeneous domains. Trans. Moscow Math. Soc. **12** (1963), 404–437.

[Wi] J. Wishart: The generalized product moment distribution in samples from a normal multivariate population. Biometrika **20A** (1928), 32–52.

[ZSSS] K.A. Zhevlakov, A.M. Slinko, I.P. Shestakov, A.I. Shirshov: Rings that are nearly associative. Academic Press, New York (1982).

Index

Biography of Max Koecher

Max Koecher was born in 1924 in Weimar. In 1947 he started studying mathematics and physics at the University of Göttingen, where he received his PhD under the guidance of Hel Braun in 1951. The following year he joined the group of Hans Petersson at the University of Münster, where he obtained his "Habilitation" in 1954.

Max Koecher stayed in Münster as "Privatdozent" and professor until 1962, when he became a full professor at the University of Munich. In 1970 he returned to Münster, where he stayed until his retirement in 1989. Max Koecher passed away in 1990 in Lengerich.

During his academic career Koecher held a number of visiting positions, in particular at the University of Minnesota in the academic year 1961/62, where these notes had their origin. Moreover he worked at Yale University in 1965/66, at the University of Aarhus in spring 1967, at Rice University in 1969 and at the University of Ottawa in 1977.

The mathematical work of Max Koecher refers to

- modular forms of several variables
- domains of positivity and Jordan algebras
- Lie algebras and Jordan algebras
- bounded symmetric domains and Jordan triple systems
- non-associative algebras.

In addition to his research papers, he wrote several influential textbooks and monographs.

For a detailed biography see the obituary [KrP].

Printing: Weihert-Druck GmbH, Darmstadt
Binding: Buchbinderei Schäffer, Grünstadt

Lecture Notes in Mathematics

For information about Vols. 1–1520
please contact your bookseller or Springer-Verlag

Vol. 1563: E. Fabes, M. Fukushima, L. Gross, C. Kenig, M. Röckner, D. W. Stroock, Dirichlet Forms. Varenna, 1992. Editors: G. Dell'Antonio, U. Mosco. VII, 245 pages. 1993.

Vol. 1564: J. Jorgenson, S. Lang, Basic Analysis of Regularized Series and Products. IX, 122 pages. 1993.

Vol. 1565: L. Boutet de Monvel, C. De Concini, C. Procesi, P. Schapira, M. Vergne. D-modules, Representation Theory, and Quantum Groups. Venezia, 1992. Editors: G. Zampieri, A. D'Agnolo. VII, 217 pages. 1993.

Vol. 1566: B. Edixhoven, J.-H. Evertse (Eds.), Diophantine Approximation and Abelian Varieties. XIII, 127 pages. 1993.

Vol. 1567: R. L. Dobrushin, S. Kusuoka, Statistical Mechanics and Fractals. VII, 98 pages. 1993.

Vol. 1568: F. Weisz, Martingale Hardy Spaces and their Application in Fourier Analysis. VIII, 217 pages. 1994.

Vol. 1569: V. Totik, Weighted Approximation with Varying Weight. VI, 117 pages. 1994.

Vol. 1570: R. deLaubenfels, Existence Families, Functional Calculi and Evolution Equations. XV, 234 pages. 1994.

Vol. 1571: S. Yu. Pilyugin, The Space of Dynamical Systems with the C^0-Topology. X, 188 pages. 1994.

Vol. 1572: L. Göttsche, Hilbert Schemes of Zero-Dimensional Subschemes of Smooth Varieties. IX, 196 pages. 1994.

Vol. 1573: V. P. Havin, N. K. Nikolski (Eds.), Linear and Complex Analysis – Problem Book 3 – Part I. XXII, 489 pages. 1994.

Vol. 1574: V. P. Havin, N. K. Nikolski (Eds.), Linear and Complex Analysis – Problem Book 3 – Part II. XXII, 507 pages. 1994.

Vol. 1575: M. Mitrea, Clifford Wavelets, Singular Integrals, and Hardy Spaces. XI, 116 pages. 1994.

Vol. 1576: K. Kitahara, Spaces of Approximating Functions with Haar-Like Conditions. X, 110 pages. 1994.

Vol. 1577: N. Obata, White Noise Calculus and Fock Space. X, 183 pages. 1994.

Vol. 1578: J. Bernstein, V. Lunts, Equivariant Sheaves and Functors. V, 139 pages. 1994.

Vol. 1579: N. Kazamaki, Continuous Exponential Martingales and *BMO*. VII, 91 pages. 1994.

Vol. 1580: M. Milman, Extrapolation and Optimal Decompositions with Applications to Analysis. XI, 161 pages. 1994.

Vol. 1581: D. Bakry, R. D. Gill, S. A. Molchanov, Lectures on Probability Theory. Editor: P. Bernard. VIII, 420 pages. 1994.

Vol. 1582: W. Balser, From Divergent Power Series to Analytic Functions. X, 108 pages. 1994.

Vol. 1583: J. Azéma, P. A. Meyer, M. Yor (Eds.), Séminaire de Probabilités XXVIII. VI, 334 pages. 1994.

Vol. 1584: M. Brokate, N. Kenmochi, I. Müller, J. F. Rodriguez, C. Verdi, Phase Transitions and Hysteresis. Montecatini Terme, 1993. Editor: A. Visintin. VII. 291 pages. 1994.

Vol. 1585: G. Frey (Ed.), On Artin's Conjecture for Odd 2-dimensional Representations. VIII, 148 pages. 1994.

Vol. 1586: R. Nillsen, Difference Spaces and Invariant Linear Forms. XII, 186 pages. 1994.

Vol. 1587: N. Xi, Representations of Affine Hecke Algebras. VIII, 137 pages. 1994.

Vol. 1588: C. Scheiderer, Real and Étale Cohomology. XXIV, 273 pages. 1994.

Vol. 1589: J. Bellissard, M. Degli Esposti, G. Forni, S. Graffi, S. Isola, J. N. Mather, Transition to Chaos in Classical and Quantum Mechanics. Montecatini Terme, 1991. Editor: 2S. Graffi. VII, 192 pages. 1994.

Vol. 1590: P. M. Soardi, Potential Theory on Infinite Networks. VIII, 187 pages. 1994.

Vol. 1591: M. Abate, G. Patrizio, Finsler Metrics – A Global Approach. IX, 180 pages. 1994.

Vol. 1592: K. W. Breitung, Asymptotic Approximations for Probability Integrals. IX, 146 pages. 1994.

Vol. 1593: J. Jorgenson & S. Lang, D. Goldfeld, Explicit Formulas for Regularized Products and Series. VIII, 154 pages. 1994.

Vol. 1594: M. Green, J. Murre, C. Voisin, Algebraic Cycles and Hodge Theory. Torino, 1993. Editors: A. Albano, F. Bardelli. VII, 275 pages. 1994.

Vol. 1595: R.D.M. Accola, Topics in the Theory of Riemann Surfaces. IX, 105 pages. 1994.

Vol. 1596: L. Heindorf, L. B. Shapiro, Nearly Projective Boolean Algebras. X, 202 pages. 1994.

Vol. 1597: B. Herzog, Kodaira-Spencer Maps in Local Algebra. XVII, 176 pages. 1994.

Vol. 1598: J. Berndt, F. Tricerri, L. Vanhecke, Generalized Heisenberg Groups and Damek-Ricci Harmonic Spaces. VIII, 125 pages. 1995.

Vol. 1599: K. Johannson, Topology and Combinatorics of 3-Manifolds. XVIII, 446 pages. 1995.

Vol. 1600: W. Narkiewicz, Polynomial Mappings. VII, 130 pages. 1995.

Vol. 1601: A. Pott, Finite Geometry and Character Theory. VII, 181 pages. 1995.

Vol. 1602: J. Winkelmann, The Classification of Three-dimensional Homogeneous Complex Manifolds. XI, 230 pages. 1995.

Vol. 1603: V. Ene, Real Functions – Current Topics. XIII, 310 pages. 1995.

Vol. 1604: A. Huber, Mixed Motives and their Realization in Derived Categories. XV, 207 pages. 1995.

Vol. 1605: L. B. Wahlbin, Superconvergence in Galerkin Finite Element Methods. XI, 166 pages. 1995.

Vol. 1606: P.-D. Liu, M. Qian, Smooth Ergodic Theory of Random Dynamical Systems. XI, 221 pages. 1995.

Vol. 1607: G. Schwarz, Hodge Decomposition – A Method for Solving Boundary Value Problems. VII, 155 pages. 1995.

Vol. 1608: P. Biane, R. Durrett. Lectures on Probability Theory. Editor: P. Bernard. VII, 210 pages. 1995.

Vol. 1609: L. Arnold, C. Jones, K. Mischaikow, G. Raugel, Dynamical Systems. Montecatini Terme, 1994. Editor: R. Johnson. VIII, 329 pages. 1995.

Vol. 1610: A. S. Üstünel, An Introduction to Analysis on Wiener Space. X, 95 pages. 1995.

Vol. 1611: N. Knarr, Translation Planes. VI, 112 pages. 1995.

Vol. 1612: W. Kühnel, Tight Polyhedral Submanifolds and Tight Triangulations. VII. 122 pages. 1995.

4. Lecture Notes are printed by photo-offset from the master-copy delivered in camera-ready form by the authors. Springer-Verlag provides technical instructions for the preparation of manuscripts. Macro packages in T_EX, L^AT_EX2e, L^AT_EX2.09 are available from Springer's web-pages at

http://www.springer.de/math/authors/b-tex.html.

Careful preparation of the manuscripts will help keep production time short and ensure satisfactory appearance of the finished book.

The actual production of a Lecture Notes volume takes approximately 12 weeks.

5. Authors receive a total of 50 free copies of their volume, but no royalties. They are entitled to a discount of 33.3 % on the price of Springer books purchase for their personal use, if ordering directly from Springer-Verlag.

Commitment to publish is made by letter of intent rather than by signing a formal contract. Springer-Verlag secures the copyright for each volume. Authors are free to reuse material contained in their LNM volumes in later publications: A brief written (or e-mail) request for formal permission is sufficient.

Addresses:

Professor F. Takens, Mathematisch Instituut,
Rijksuniversiteit Groningen, Postbus 800,
9700 AV Groningen, The Netherlands
E-mail: F.Takens@math.rug.nl

Professor B. Teissier, DMI, École Normale Supérieure
45, rue d'Ulm,
F-7500 Paris, France
E-mail: Teissier@ens.fr

Springer-Verlag, Mathematics Editorial, Tiergartenstr. 17,
D-69121 Heidelberg, Germany,
Tel.: *49 (6221) 487-701
Fax: *49 (6221) 487-355
E-mail: C.Byrne@Springer.de